# Physics Research and Technology

# Physics Research and Technology

**Second Harmonic Generation: Pathways of Nonlinear Photonics**
Abdel-Baset M. A. Ibrahim, PhD
Pankaj Kumar Choudhury, PhD
2022. ISBN: 978-1-68507-888-1 (Hardcover)
2022. ISBN: 979-8-88697-002-9 (eBook)

**Advances in Sustainable Materials and Technology**
Abhishek Kanoungo, PhD
Sandeep Singh, PhD
Er. Shristi Kanoungo, M.E
Ajay Goyal, PhD
2022. ISBN: 978-1-68507-967-3 (Hardcover)
2022. ISBN: 979-8-88697-070-8 (eBook)

**Quantum Field Theory and Applications**
Natale Palerma (Editor)
2022. ISBN: 978-1-68507-930-7 (Hardcover)
2022. ISBN: 978-1-68507-957-4 (eBook)

**2D Metallic Transition Metal Dichalcogenides: Fundamentals and Applications**
Brahmananda Chakraborty, PhD, Chandra Sekhar Rout, PhD (Editors)
2022. ISBN: 978-1-68507-965-9 (Hardcover)
2022. ISBN: 979-8-88697-067-8 (eBook)

**Spatial Homeostasis, Quantum Information Channel, and the Nature of Living Things Within the Framework of the Theory of Byuon**
Yuriy Baurov, PhD
2022. ISBN: 978-1-68507-590-3 (eBook)

More information about this series can be found at
https://novapublishers.com/product-category/series/physics-research-and-technology/

**Manuel B. Hutchinson**
Editor

# Electromagnetic Waves

**Advances in Applications and Research**

Copyright © 2022 by Nova Science Publishers, Inc.

**All rights reserved.** No part of this book may be reproduced, stored in a retrieval system or transmitted in any form or by any means: electronic, electrostatic, magnetic, tape, mechanical photocopying, recording or otherwise without the written permission of the Publisher.

We have partnered with Copyright Clearance Center to make it easy for you to obtain permissions to reuse content from this publication. Simply navigate to this publication's page on Nova's website and locate the "Get Permission" button below the title description. This button is linked directly to the title's permission page on copyright.com. Alternatively, you can visit copyright.com and search by title, ISBN, or ISSN.

For further questions about using the service on copyright.com, please contact:
Copyright Clearance Center
Phone: +1-(978) 750-8400     Fax: +1-(978) 750-4470     E-mail: info@copyright.com

## NOTICE TO THE READER

The Publisher has taken reasonable care in the preparation of this book, but makes no expressed or implied warranty of any kind and assumes no responsibility for any errors or omissions. No liability is assumed for incidental or consequential damages in connection with or arising out of information contained in this book. The Publisher shall not be liable for any special, consequential, or exemplary damages resulting, in whole or in part, from the readers' use of, or reliance upon, this material. Any parts of this book based on government reports are so indicated and copyright is claimed for those parts to the extent applicable to compilations of such works.

Independent verification should be sought for any data, advice or recommendations contained in this book. In addition, no responsibility is assumed by the Publisher for any injury and/or damage to persons or property arising from any methods, products, instructions, ideas or otherwise contained in this publication.

This publication is designed to provide accurate and authoritative information with regard to the subject matter covered herein. It is sold with the clear understanding that the Publisher is not engaged in rendering legal or any other professional services. If legal or any other expert assistance is required, the services of a competent person should be sought. FROM A DECLARATION OF PARTICIPANTS JOINTLY ADOPTED BY A COMMITTEE OF THE AMERICAN BAR ASSOCIATION AND A COMMITTEE OF PUBLISHERS.

Additional color graphics may be available in the e-book version of this book.

## Library of Congress Cataloging-in-Publication Data

ISBN: 979-8-88697-254-2

*Published by Nova Science Publishers, Inc. † New York*

# Contents

**Preface** ........................................................................... vii

**Chapter 1** **Microwave Absorption Properties of CoFe$_2$O$_4$-Based Materials: A Review** ................................ 1
Nguyen Quy Tuan, Chu Thi Anh Xuan and Ngo Tran

**Chapter 2** **Strong Resonance Effects in Ordered Layered Photonic Structures for Filtering, Collimation, Metrology and Spectroscopy** .......................... 29
E. Ya. Glushko

**Chapter 3** **Antenna Array of Upgraded Quadruple Sector Emitters for Wireless 4G Networks** ................. 55
Andrii Karpenko

**Chapter 4** **On the Scattering of Electromagnetic Waves by Bodies with Non-Coordinate Boundaries** ........ 67
I. E. Pleshchinskaya and N. B. Pleshchinskii

**Bibliography** ................................................................. 99

**Index** ............................................................................. 167

# Preface

Electromagnetic waves are waves that are formed due to vibrations between an electric field and a magnetic field. This book includes four chapters that present new applications and research on electromagnetic waves. Chapter One reviews the microwave absorption properties of CoFe2O4-based materials. Chapter Two examines strong resonance effects in ordered layered photonic structures for filtering, collimation, metrology and spectroscopy. Chapter Three describes an antenna array of upgraded quadruple sector emitters for wireless 4G networks. Lastly, Chapter Four explores the possibility of numerical solving scattering problems of electromagnetic waves by bodies with non-coordinate boundaries by reducing to infinite sets of linear algebraic equations relative to the coefficients of expansion of the unknown field in terms of eigen waves of coordinate domains in different coordinate systems.

Chapter 1 - The widespread use of electronic and wireless devices causes a global environmental pollutant due to electromagnetic radiation. Therefore, finding desirable microwave absorbers (MAs) is an urgent task. Spinel ferrite (AFe2O4) was one of the first MA generations due to its good magnetic loss performance and being eco-friendly and cost-effective. Along with its advantages, spinel ferrite has also shown some disadvantages (low natural resonance frequency, lack of conduction loss, and high density), which prevent it from becoming a promising MA. Cobalt spinel ferrite (CoFe2O4, CFO) contains all of the benefits and drawbacks of AFe2O4 for MA. In order to overcome these drawbacks, much effort has been put into fabricating CFO-based materials with high values of conduction loss, being lightweight, and suitable natural resonance frequency. This chapter will conduct a systematic review of methods for enhancing the microwave absorption properties of CFO-based materials. This chapter will review all simple and complex methods to achieve a desirable CFO-based microwave absorber. The simple methods could be considered as fabricating hierarchical 3D morphologies and transitional/rare-earth metal doping on the A and B sites of the CFO. The complicated method composites CFO with other magnetic materials, such as iron oxide, alloys, soft spinel ferrite, M-type hexaferrite, and garnet ferrite. The more complex method could be fabricating a composite of magnetic material (CFO) with dielectric materials (carbon, graphene, reduced graphene oxide (rGO), and MXene) components. Some studies even reported the fabrication of CFO-based absorbers by combining the methods described above, such as decorating carbon and rGO on the surface of CFO and decorating carbon/graphite/rGO on the surface of CFO/alloys composite to achieve good values of reflection loss and effective absorption bandwidth. Each method's strengths and weaknesses will also be mentioned in this chapter.

Chapter 2 - A planar photonic resonator containing unitary defect in the middle of the structure can exhibit a system of extraordinary narrow resonance peaks of transmission on the background of perfect reflection. The properties of standing modes inside the polyethylene (polypropylene)/

silicon plane resonators in the total intrinsic reflection region and unusual manifestations of THz transmission spectra in centimeter and millimeter wavelength range were studied. It is shown that the angle and frequency half-widths of the resonance peak can be less than 10-9 of the magnitude of angle and frequency in dependence on the number of periods. This allows one to form collimated beams with the divergence measured in a fraction of a microdegree. It is shown that a plane resonator containing a central defect transforms the frequency divided peaks into the outgoing-transmitted beams of various directions like a prism transforms light. This opens the way for precision measurements of angle and frequency distribution of THz radiation. It is proposed to use the existing extremely sharp peaks of transmission in planar resonators containing a central defect for aims of spectroscopy and metrology. A new spectroscopy technique is proposed based on the existing sharp transmission resonances using the conception of accumulating reservoir of electromagnetic field. Extra-high extent of collimation resulting from the usage of defected photonic resonators gives an opportunity to form long and stable channels of communication in the THz frequency range.

Chapter 3 - Based on the "Quados" radiators, developed an optimized antenna array with a gain of 17÷12,5 dBi in the frequency band 3 (1710÷1880 GHz), designed for efficient operation with 4G LTE-1800 mobile operators in areas of weak signal strength. The following minimum values were selected as optimization criteria: 1. Standing wave coefficient in the operating frequency range; 2. Weight and sail of the antenna; 3. The number of matching devices.

Chapter 4 - This chapter explores the possibility of numerical solving scattering problems of electromagnetic wave by bodies with non-coordinate boundaries by reducing to infinite sets of linear algebraic equations relative to the coefficients of expansion of the unknown field in terms of eigen waves of coordinate domains in different coordinate systems. In Cartesian coordinates scattering problems of an electromagnetic wave by a non-coordinate media interface and by a conductive thin screen in a plane waveguide are considered. Two-dimensional and three-dimensional problems of diffraction of a plane wave by a periodically perturbed media interface in the open space are considered also. Expansions of the unknown field in terms of cylindrical and spherical waves are used in solving the problem of diffraction of an electromagnetic wave by a cylindrical rod with a noncoordinate boundary and a two-dimensional wave diffraction problem by the axis-symmetrical body. Some results of computational experiments are given.

# Chapter 1

# Microwave Absorption Properties of $CoFe_2O_4$-Based Materials: A Review

## Nguyen Quy Tuan[1], PhD, Chu Thi Anh Xuan[2], PhD and Ngo Tran[3,4,*], PhD

[1]The University of Danang - University of Science and Education, Da Nang, Vietnam
[2]Institute of Science and Technology, TNU-University of Science, Thai Nguyen, Vietnam
[3]Institute of Research and Development, Duy Tan University, Da Nang, Vietnam
[4]Faculty of Natural Sciences, Duy Tan University, Da Nang, Vietnam

## Abstract

The widespread use of electronic and wireless devices causes a global environmental pollutant due to electromagnetic radiation. Therefore, finding desirable microwave absorbers (MAs) is an urgent task. Spinel ferrite ($AFe_2O_4$) was one of the first MA generations due to its good magnetic loss performance and being eco-friendly and cost-effective. Along with its advantages, spinel ferrite has also shown some disadvantages (low natural resonance frequency, lack of conduction loss, and high density), which prevent it from becoming a promising MA. Cobalt spinel ferrite ($CoFe_2O_4$, CFO) contains all of the benefits and drawbacks of $AFe_2O_4$ for MA. In order to overcome these drawbacks, much effort has been put into fabricating CFO-based materials with high values of conduction loss, being lightweight, and suitable natural resonance frequency. This chapter will conduct a systematic review of methods for enhancing the microwave absorption properties of CFO-based materials. This chapter will review all simple and complex methods to achieve a desirable CFO-based microwave absorber. The simple methods could be considered as fabricating hierarchical 3D

---

[*] Corresponding Author's Email: tranngo@duytan.edu.vn.

In: Electromagnetic Waves
Editor: Manuel B. Hutchinson
ISBN: 979-8-88697-254-2
© 2022 Nova Science Publishers, Inc.

morphologies and transitional/rare-earth metal doping on the A and B sites of the CFO. The complicated method composites CFO with other magnetic materials, such as iron oxide, alloys, soft spinel ferrite, M-type hexaferrite, and garnet ferrite. The more complex method could be fabricating a composite of magnetic material (CFO) with dielectric materials (carbon, graphene, reduced graphene oxide (rGO), and MXene) components. Some studies even reported the fabrication of CFO-based absorbers by combining the methods described above, such as decorating carbon and rGO on the surface of CFO and decorating carbon/graphite/rGO on the surface of CFO/alloys composite to achieve good values of reflection loss and effective absorption bandwidth. Each method's strengths and weaknesses will also be mentioned in this chapter.

**Keywords**: ferrite, $CoFe_2O_4$, electromagnetic properties, microwave absorption properties, electromagnetic interference shielding effectiveness

## Introduction

In modern society, many inevitable pollutants badly affect daily life and the human body [1-3]. Among these pollutants, electromagnetic (EM) pollution can be divided into two types: biological pollution cause by EM wave interference and EM waves radiation emitted by electronic devices [4-6]. EM pollution is more difficult to control than other types of pollution, such as air and water pollution. It is worth noticing that the EM wave brought many advantages to human life. Therefore, humans should learn to control the harms of EM pollution while enjoying its advantages. In order to do this, many researchers are attempting to chase excellent EM wave absorbing materials, which should be excellent in absorption, wideband, low density, thin thickness, etc.

With its advantages, such as strong magnetic loss ability and low cost, ferrite was considered the first EM wave-absorbing material. Among ferrite materials, cobalt spinel ferrite ($CoFe_2O_4$, CFO) has been considered a promising EM wave absorbing material, especially in the gigahertz (GHz) range. However, some disadvantages, such as high density, lack of conduction loss, and low ferromagnetic resonance frequency ($f_{FMR}$), prevented CFO from becoming an excellent microwave absorbing material. In order to resolve these issues, much effort has been put into improving the microwave absorption properties of CFO-based materials. Some methods could be as follows: fabricating unique morphologies, [7] doping on Co and Fe sites, [8, 9] compositing with $Fe_2O_3$, [10] and magnetic materials (metals/alloys, [11-13] other spinel ferrites, [14] hexaferrite, [15] and garnet [16]). Compositing with allotropies of carbon was also used to enhance the microwave dissipation features of CFO-based materials: with carbon, [17-23] graphene, [24, 25] graphene oxide (GO), reduced GO (rGO), [26-29] and even conducting polymer and MXene [30]. In order to remarkably enhance the microwave absorption properties, ternary composites based on CFO were also fabricated: CFO/C/rGO, [31] composites of CFO with metals/alloys and carbon/graphite/rGO [32-35].

To gain an understanding of the methods for improving the microwave absorption properties of CFO-based materials, microwave absorption data from these materials was collected, summarized, and presented in this book chapter.

## Microwave Absorption of Bare CFO

As mentioned in the Introduction, CFO has been considered a good microwave absorber due to its high saturation magnetization and large magnetocrystalline anisotropy, leading to its large complex permeability [23, 27]. However, CFO also has some unbeneficial characteristics (such as high density, low $f_{FMR}$, lack of conduction loss, etc.) that prevent CFO from becoming an excellent microwave absorber. In order to overcome the high density of CFO, Ni et al. fabricated a hollow CFO and then investigated its microwave absorbing performances, which achieved a reflection loss (RL) of –15.23 dB at $f$ = 16.98 GHz and an effective absorption bandwidth (EAB) of 2.72 GHz (from 15.28 to 18 GHz) for $t$ = 5.5 mm [36]. Sang et al. reported that mesoporous CFO could reach a minimum RL of –27.36 dB with $t$ = 3 mm [37]. The improvement of the microwave absorption performance of mesoporous CFO could mainly originate from multiple reflection EM waves by the mesoporous structure combined with high EM wave attenuation and good impedance matching. Jiang et al. reported that the fabrication of CFO with flower-like and crochet ball-like morphologies could improve CFO microwave absorption [38]. The flower-like CFO achieved an RL of –23 dB and an EAB of 13 GHz, while these values for the crochet ball-like CFO were –40 dB and 12.5 GHz, respectively. Jafarian et al. reported that brain-coral- and rod-like CFO could increase the microwave absorption properties, which reached an RL of –31.2

dB at $f = 8.1$ GHz for $t = 10$ mm and $-15$ dB at $f = 12.4$ GHz for $t = 9$ mm, respectively [39]. These reports showed pretty low values of RL and narrow values of EAB due to the lack of conduction loss. In order to overcome CFO's drawbacks, doping and compositing are two of the most frequently used methods.

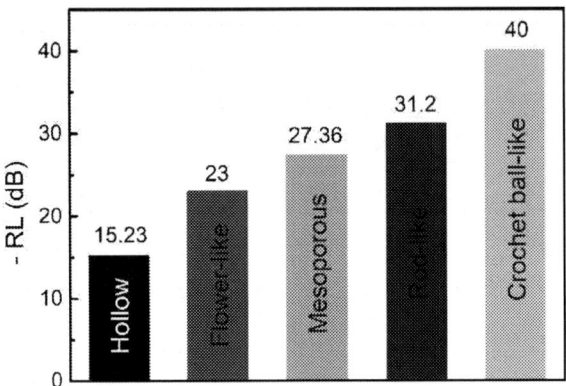

**Figure 1.** Comparison of RL values for CFO with different morphologies.

## Microwave Absorption of Doped CFO and Composites of CFO with Other Spinel Ferrites

In the case of doping, transition metals doped to the A- and/or B-sites of CFO were mainly studied. Nickel doped to the Co site of the CFO could improve RL up to $-37.66$ dB and EAB of 2.64 GHz for doping concentration of $x = 0.1$ ($Ni_{0.1}Co_{0.9}Fe_2O_4$) and the thickness of 2.1 mm [40]. Besides transition metals, rare-earth metals could be used to dope to the Fe sites of CFO to enhance the microwave absorption performance. La-doped to Fe site of CFO ($CoLa_{0.12}Fe_{1.88}O_4$) could reach a pretty high RL of $-46.47$ dB and EAB of 4.9 GHz for $t = 3$ mm [37]. The doping of La assisted in the increased specific surface area, imaginary parts of complex permeability and complex permittivity, leading to increased microwave absorption performances. Besides doping to the Co and Fe sites of CFO, compositing with other spinel ferrites could be the method to increase the microwave absorption performance of CFO.

A composite of CFO and $NiFe_2O_4$ (NFO) with core/shell structure could achieve RL of $-20.1$ dB at $f = 9.7$ GHz and EAB of 4 GHz for $t = 4.5$ mm [41].

As we can see, doping to the Co or Fe site of CFO improved microwave absorption performance more than the composite of CFO/NFO, as shown in Figure 2. This could be attributed to generating free ions by the doing, leading to improving dielectric loss.

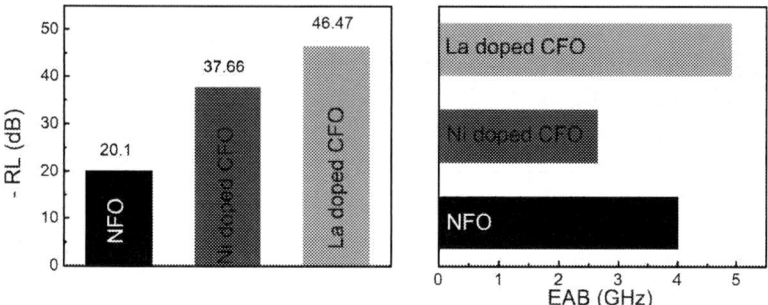

**Figure 2.** Comparison of RL and EAB values for Ni and La-doped CFO and CFO/NFO composite.

## Microwave Absorption of Composites CFO with Metals or Alloys

Mahdikhah et al. reported that the CFO/Fe nanocomposite could reach the maximum RL of −27 dB at a frequency of 11.2 GHz with 30 wt% of Fe in the composite [42]. The enhancement in microwave absorption performance of the CFO/Fe nanocomposite compared to that of the CFO could be attributed to a decrease in particle size and increased specific surface area.

FeCo/CFO composite powder with 17wt% of FeCo embedded in paraffin exhibited a pretty good RL of −20 dB and −15 dB in the C and X bands for the matching thicknesses of 9 and 6.7 mm, respectively [43]. With the same composite composition, Liu et al. reported a higher value of RL of −52.14 dB at 12.34 GHz with a thin thickness of 1.7 mm [44]. Zhou et al. synthesized composites with a composition of $(1-x)$CFO/$x$CoFe with $x$ = 0.3, 0.4, 0.5, and 0.6. Among the composites, the $x$ = 0.6 (0.4CFO/0.6Co$_4$Fe$_6$) composite showed the best microwave absorption performance with RL = − 58.22 dB at 12.96 GHz and EAB = 4.16 GHz with $t$ = 1.45 mm [45]. The microwave absorption performances of the composites of CFO with metals or alloys depend on the concentration of the metals or alloys in the composites, where RL values seem to increase with the metals or alloys.

## Microwave Absorption of Composites CFO with Hexaferrite

The porous hollow of CFO/BaFe$_{12}$O$_{19}$ (BaM) microrods could reach an RL of –25 dB at a frequency of 13 GHz and a pretty large EAB of 4.1 GHz (from 8.2 to 16.3 GHz) [46]. In contrast to the CFO/BaM composite, the SrFe$_{12}$O$_{19}$ (SrM)/CFO core/shell nanocomposite showed bad microwave absorption performance with an RL = –2.5 dB [47]. However, with a different fabrication method, Tyagi et al. could turn the RL value of SrM/CFO nanocomposite up to –27.6 dB at $f$ = 10.8 GHz [48]. The microwave absorption performance could be significantly enhanced when fabricated into a 3D structure for CFO and SrM. Composites of brain-coral-like structures CFO and SrM could achieve an RL value of –38 dB and an EAB of 2.6 GHz for $t$ = 2.5 mm [39]. The microwave dissipation feature boost could be assigned to the hierarchical configuration, leading to electromagnetic loss features. Overall, the composites of CFO with BaM and SrM were strongly related to the morphology of BaM and SrM, where samples with high porosity could achieve better microwave absorption performances.

Besides M-type, W-type hexaferrite could also be used as a component to composite with CFO to improve the microwave absorption performance, where SrZnCoFe$_{16}$O$_{27}$/CFO could reach an RL value of –8.1 dB at 10.1 GHz [49].

## Microwave Absorption of Composites CFO with Perovskite and Iron Oxide

A composite of NiTiO$_3$ and CFO with a ratio of NiTiO$_3$:CFO = 1:3 could reach an RL of –21.7 dB at 17.1 GHz for $t$ = 10 mm [50]. The microwave attenuation by the composite could be attributed to magnetic loss by increasing exchange resonance. Besides perovskite, iron(III) oxide, Fe$_2$O$_3$, could also be used as a component to tune the microwave absorption of CFO. Flake-like composite of the Fe$_2$O$_3$/CFO could reach RL of –60 dB at 16.5 mm and EAB of 5 GHz (from 13 to 18 GHz) for $t$ = 2 mm [51]. The enhancement in microwave absorption of Fe$_2$O$_3$/CFO composite could have resulted from the porous core-shell structure, electron polarization between Co and Fe, and balancing dielectric and magnetic losses.

## Microwave Absorption of Composites CFO with Carbon

As mentioned in the above sections, the microwave absorption properties of CFO could be significantly modified by fabricating with different morphologies, doping, and compositing with other magnetic materials. In these cases, the attenuation of microwave energy mostly comes from magnetic loss. In order to improve microwave absorption properties of CFO-based materials by enhancing dielectric loss, many groups have reported compositing with carbon and its allotropes.

Salimkhani et al. reported that 3D carbon fiber and CFO composite could reach RL of – 10.25 dB at 11.39 GHz [52]. Zhou et al. reported that the composite of colloidal carbon spheres (CCSs) with hollow spheres CFO showed RL of –12.6 dB and EAB of 1.4 GHz for $t$ = 6.7 mm [53]. The attenuation of the microwaves originated from their reflections and scattering in the absorbers combined with interfacial polarization and eddy current loss.

Interestingly, CFO coated on superhydrophobic wood showed good microwave absorption performance with an RL of –12.3 dB at $f$ = 16 GHz, originating from the dielectric loss [54]. This work provides a route for designing an outdoor wood that could absorb microwave energy at a specific frequency.

Composite of CFO and mesoporous carbon nanofibers (C NFs) could reach RL of –14.0 dB at 9 GHz for $t$ = 3.5 mm [55]. At the same time, $t$ = 2.5 mm could reach a maximum EAB value of 3.6 GHz. Composite 3D carbon and CFO could reach RL of –19.5 dB at $f$ = 15.45 GHz and EAB of 6.63 GHz for $t$ = 2.5 mm [56]. The good microwave absorption performance could be attributed to the strong attenuation and good impedance matching.

Li et al. reported that a porous carbon ball/CFO composite could reach an RL of –37.2 dB and an EAB of 4.2 GHz (from 12.9 to 17.1 GHz) for $t$ = 1.5 mm and 10 wt% filler loading [19]. The excellent microwave absorption resulted from improved electrical conductivity by the porous carbon ball and high electromagnetic loss. A composite of carbon porous derived from the eggshell membrane and CFO could reach an RL of –49.6 dB with 30 wt% of filler [17]. The good value of RL could be attributed to the good attenuation ability and impedance matching conditions, which resulted from the high porosity structure and significant synergetic effects from the magnetic and dielectric parts of the composite.

Xu et al. reported that rod-like porous carbon derived from cotton composited with CFO NPs improved the microwave absorption performance of CFO-based materials with an RL of –48.2 dB and an EAB of 4.8 GHz for $t$

= 2.5 mm [57]. The microwave absorption enhancement could have resulted from a conductive network supported by rod-like porous carbon and interfacial polarization from massive heterointerfaces.

**Table 1.** Comparison of microwave absorption properties of CFO/carbon composites

| Composite | Thickness (mm) | RL (dB) | EAB (GHz) | Refs. |
|---|---|---|---|---|
| CFO/carbon fibers | - | −10.25 | - | [52] |
| CFO/carbon nanofibers | 3.5 and 2.5 | −14.0 | - | [55] |
| CFO/colloidal carbon spheres | 6.4 | −12.6 | - | [53] |
| CFO/superhydrophobic wood | - | −12.3 | - | [54] |
| CFO/porous carbon ball | 1.5 | −37.2 | 4.2 | [19] |
| CFO/porous carbon (eggshell membrane) | - | −49.6 | 4.2 | [17] |
| CFO/rod-like porous carbon | 2.5 | −48.2 | 4.8 | [57] |
| CFO/porous carbon nanosheets | 2 | −52.29 | 5.36 | [58] |
| CFO/hierarchical carbon (onion skin) | 2.46 and 1.70 | −53.54 | 4.92 | [59] |

\* For composites with two thickness values: the first number is the thickness for achieving the maximum RL value, while the second is for achieving the maximum EAB value.

In another report, the composite of CFO with another allotrope of carbon as porous carbon nanosheet could significantly increase the RL to −52.29 dB for a thin $t$ of 2 mm and an EAB of 5.36 GHz (12.64–18 GHz) with only 20 wt% filler loading [58]. The strong microwave attenuation capability and excellent impedance matching could result from a 3D hierarchical porous structure, leading to weak surface reflection, strong dissipation capacity, and multiple reflection loss. Hierarchical carbon derived from onion-skin composited with hollow CFO could reach RL of −53.54 dB for $t$ = 2.46 mm and EAB of 4.92 GHz for $t$ = 1.7 mm [59]. This composite could achieve strong multi-band absorption, which could be attributed to conduction, polarization, magnetic losses, and impedance matching.

Overall, the microwave absorption performances of CFO/carbon composites are strongly related to the morphology of CFO and carbon. Composites with hierarchical structures with high porosity of CFO and carbon could achieve excellent microwave absorption performances. A detailed comparison of their RL and EAB values is listed in Table 1.

## Microwave Absorption of Composites CFO with Carbon Nanotubes (CNTs)

Feng et al. reported that a composite of coiled carbon nanotubes (CCNTs) and CFO could achieve an RL of −14 dB and an EAB of 4 GHz [60]. The microwave attenuation by CCNTs/CFO originated from the conductive loss and dielectric loss, which resulted from natural resonance, exchange resonance, eddy current loss, interfacial polarization, and relaxation effects.

Zhang et al. reported that the microwave absorption performance of CFO/CNTs nanocomposites prepared by the solvothermal method could be modulated by annealing temperature. CFO/CNTs annealed at 260°C could reach an RL of −15.7 dB and an EAB of 2.5 GHz, which was the best compared to the others in the study [61].

Composite of single-walled carbon nanotubes (SWCNTs) and CFO could reach an RL of −30.7 dB at 12.9 GHz and a large EAB of 7.2 GHz for $t = 2$ mm with only 10 wt% filler loading [62]. The good microwave absorption properties could be derived from the complex multi-dipole polarization process, impedance matching, and synergetic effect of the multiple dielectric and magnetic losses. Yuan et al. reported that CFO/CNTs nanocomposite could reach RL of −37.39 dB and EAB of 5.2 GHz for $t = 1.7$ mm [63].

While hollow spherical CFO composited with CNTs using the CVD method could reach RL of −32.8 dB and EAB of 5.7 GHz for $t = 2$ mm [64]. This study provided a method to achieve a lightweight microwave absorbing material. CNTs coated CFO composite prepared using an analogous metal-organic framework (MOF) precursor could reach RL of −34.6 dB at $f = 13.4$ GHz and EAB of 7.1 GHz (10.0–11.7 GHz) for $t = 2.5$ mm [65]. The increase in microwave absorption properties originated from the significantly increased dielectric loss and improved dissipation of magnetic energy and its cage structure.

Ashfaq et al. reported that a composite of CFO and multi-walled CNTs (MWCNTs) could reach intensive an RL of −50.8 dB for $t = 4.2$ mm and an EAB of 3.36 GHz for $t = 1.6$ mm [66]. Enhancement of the composite resulted from a supplement of MWCNTs content and other factors such as dielectric loss, magnetic loss, conduction loss, impedance matching, interfacial and dipole polarization, and EM attenuation.

Similar to composites of CFO/carbon, microwave absorption performance of composites of CFO with CNTs was strongly related to the morphology of CNTs, while the CFO/MWCNTs composite could achieve the best microwave absorption dissipation features. A detailed comparison of their RL and EAB values is listed in Table 2.

Table 2. Comparison of microwave absorption properties of CFO/CNTs composites

| Composite | Thickness (mm) | RL (dB) | EAB (GHz) | Refs. |
|---|---|---|---|---|
| CFO/coiled CNTs | - | −14 | 4 | [60] |
| CFO/CNTs nanocomposites | - | −15.7 | 2.5 | [61] |
| CFO/CNTs nanocomposites | 1.7 | −37.39 | 5.2 | [63] |
| CFO/SWCNTs | 2 | −30.7 | 7.2 | [62] |
| Hollow spherical CFO/CNTs | 2 | −32.8 | 5.7 | [64] |
| CFO/CNTs (using analogous MOF precursor) | 2.5 | −34.6 | 7.1 | [65] |
| CFO/MWCNTs | 4.2 | −50.8 | 3.36 | [66] |

## Microwave Absorption of Composites CFO with Graphite

Yadav et al. reported that a composite of CFO NPs, exfoliated nano graphite (NG) embedded in polymethylmethacrylate (PMMA) could reach an RL of −36.7 dB and an EAB of 7.1 GHz [67]. Attenuation of the microwave by this composite originated from significant magnetic and dielectric losses.

## Microwave Absorption of Composites CFO with Graphene

Fu et al. reported that the CFO hollow sphere/graphene composite could reach an RL of −18.5 dB at 12.9 GHz and an EAB of 3.7 GHz (from 11.3 to 15.0 GHz) for $t = 2$ mm [68]. The good microwave dissipation features could be attributed to good electrical and magnetic properties. Sandwich-structured CFO/graphene nanocomposite could reach RL of −36.4 dB at 12.9 GHz for $t = 2.5$ mm [69]. The good microwave dissipation features could be attributed to the synergistic effects of dielectric loss (from graphene) and magnetic loss (from CFO).

Zhang et al. reported that rugby-shaped CFO/graphene could achieve an RL of −39 dB at $f = 10.9$ GHz and an EAB of 4.7 GHz (9.6–14.3 GHz) for $t = 2$ mm [70]. This study proved that the microwave absorption properties of the rugby-shaped CFO/graphene composite were better than those of the CFO NPs/graphene. A composite of flower-like CFO and graphene could reached an RL of −42 dB at $f$ = 12.9 GHz and an EAB of 4.59 GHz (11.2–15.79 GHz) for $t$ = 2 mm [24]. These values of RL and EAB proved that CFO/graphene has excellent microwave absorption performance with low density for practical applications.

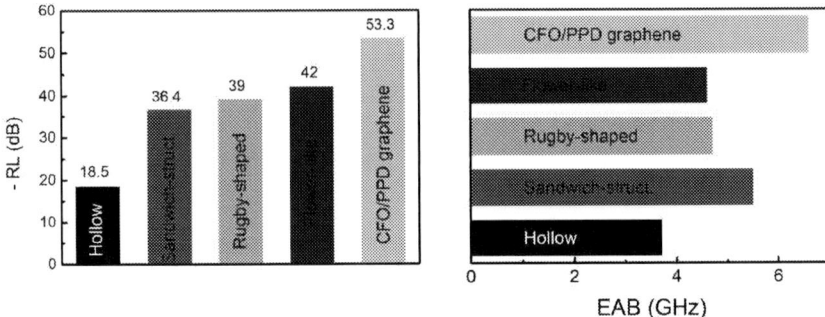

**Figure 3.** Comparison of RL and EAB values for CFO/graphene composites.

Ma et al. reported that the composite of *p*-phenylenediamine (PPD) functionalized graphene (PG) and CFO could be achieved with an RL of −53.3 dB and an EAB of 6.6 GHz (11.4–18 GHz) for $t$ = 2.4 mm [71]. The good microwave absorption PG/CFO could have originated from the hierarchical porous structure and synergistic effects of magnetic loss and dielectric loss.

Similar to other allotropes of carbon, the composites of CFO and graphene could reach excellent microwave absorption performances for high porosity and having hierarchical structure samples. As shown in Figure 3, RL and EAB values strongly depend on the morphology and structure of composites, while absorber thickness varied in the range of 2–2.5 mm.

## Microwave Absorption of Composites CFO with rGO

Guan et al. reported that a composite of CFO nanorods and rGO with a weight ratio of rGO of 20% could reach an RL of −11.1 dB at $f$ = 17 GHz and an EAB of 2.3 GHz for $t$ = 2.5 mm [72]. Moitra et al. reported that the nanocomposite of CFO-rGO prepared by *in-situ* co-precipitation with 85 wt% CFO and 15 wt% rGO could achieve a reasonably good microwave absorption performance with an RL of −31.31 dB at 9.05 GHz and an EAB of 2.72 GHz (from 8.2 to 10.92 GHz) [73]. The hetero-architectural structure mainly contributed to the microwave absorption performance of the 85CFO/15rGO composite.

Liu et al. reported that CFO/rGO nanocomposites could reach RL of $-40$ dB at $f = 6.8$ GHz and EAB of 2.7 GHz (5.8–8.5 GHz) for $t = 4$ mm [74]. The good microwave absorption properties of the CFO/rGO composite in this work could be derived from the balance between dielectric loss and magnetic loss, an increase in interfacial polarization, and polarization relaxation.

Liu et al. reported that CFO/rGO with different porous structures could reach an RL value of $-43.0$ dB at 14.4 GHz and an EAB of 5.6 GHz for $t = 1.8$ mm [75]. These good values originated from the rGO porous structure, the cooperation of magnetic and dielectric materials, and interfacial polarization.

A composite of rGO/CFO synthesized by a facile method could reach an RL of $-44.1$ dB at 15.6 GHz for $t = 1.6$ mm, while the widest EAB was obtained as 4.7 GHz (13.3–18.0 GHz) for $t = 1.5$ mm [27]. Zong et al. reported a rGO/CFO composite showed an intensive RL of $-47.9$ dB at 12.4 GHz and a wide EAB of 5 GHz for $t = 2.3$ mm [76].

In another report, an rGO/CFO composite with nanorod morphology of CFO could reach an RL value of $-56.3$ dB at $f = 16.9$ GHz for only $t = 1.4$ mm. While the widest EAB of 4.3 GHz could be reached for a thickness of 1.6 mm [77]. This composite could be applied as a strong capacity and lightweight microwave absorbing material. Liu et al. reported that hierarchical CFO/rGO porous nanocomposites could reach an RL of $-57.7$ dB and a wide EAB of 5.8 GHz (8.3–14.1 GHz) for $t = 2.8$ mm [28]. The impressive microwave absorption performance of CFO/rGO originated from impedance matching and attenuation ability enhancement.

Table 3. Comparison of microwave absorption properties of CFO/rGO composites

| Composite | Thickness (mm) | RL (dB) | EAB (GHz) | Refs. |
|---|---|---|---|---|
| CFO nanorods/rGO | 2.5 | −11.1 | - | [72] |
| 85 wt% CFO/15 wt% rGO | - | −31.31 | 2.72 | [73] |
| CFO/rGO nanocomposites | 4.0 | −40.0 | 2.7 | [74] |
| CFO/rGO porous structure | 1.8 | −43.0 | 5.6 | [75] |
| CFO/rGO (facile method) | 1.6 | −44.1 | 4.7 | [27] |
| CFO/rGO | 2.3 | −47.9 | 5.0 | [76] |
| Nanorods CFO/rGO | 1.4 | −56.3 | 4.3 | [77] |
| Hierarchical CFO/rGO porous | 2.8 | −57.7 | 5.5 | [28] |
| CFO/rGO nanocomposite | 2.1 | −67.58 | 6.3 | [78] |

Another group reported that CFO/rGO nanocomposites could reach an intensive RL value of –67.58 dB and a broad EAB of 6.3 GHz for $t = 2.1$ mm and only 10% filler loading [78]. The enhancement of microwave dissipation features in CFO/rGO nanocomposites originated from the synergistic effect of hybrid microstructure, impedance matching, and various loss mechanisms.

Similar to other allotropes of carbon above, the composites of CFO and rGO could achieve excellent microwave absorption performances for composites having hybrid microstructures and having various loss mechanisms. A detailed comparison of their RL and EAB values is listed in Table 3.

## Microwave Absorption of Composites CFO with MXene

$Ti_3C_2$ MXene is a potential microwave absorbing material. However, its poor impedance matching and lack of magnetic loss prevent $Ti_3C_2$ MXene from becoming a good microwave absorber. Therefore, $Ti_3C_2$ MXene could be composited with CFO to improve the microwave absorption. He et al. reported that CFO-$Ti_3C_2$ could achieve a good RL of –30.9 dB and an EAB of 8.5 GHz for a $t$ of only 1.5 mm [79]. The good microwave absorption properties of CFO-$Ti_3C_2$ came from the increased percolation threshold, controlling complex permittivity, and the appearance of CFO.

## Microwave Absorption of Composites CFO with Conducting Polymer

Ismail et al. reported that the composite of CFO with polyaniline (PANI) doped with para toluene sulphonic acid (PTSA) could reach an RL of –28.4 dB at 8.1 GHz [80]. Another group reported that CFO/PANI could reach RL of –41.9 dB at 15.8 GHz for $t = 1.9$ mm [81]. The excellent originated from a robust dielectric loss at the interface of CFO and PANI.

Besides PANI, polypyrrole (PPy) could be used as a component in the CFO-based composites. Ren et al. reported that hierarchical CFO@PPy hollow nanocubes could reach an RL of –43.85 dB at 9.7 GHz for $t = 3$ mm [82]. At the same time, the largest EAB was 5.7 GHz for $t = 2$ mm. The enhancement of microwave absorption performance for CFO@PPy originated from multiple loss mechanisms and impedance matching improvement.

Instead of prepared CFO/PPy composite, Zhang et al. prepared a bi-layer absorber with PPy as the upper layer and CFO as the lower layer for a thickness of 2.3 mm. The CFO/PPy with CFO and PPy thicknesses of 1 and 1.3 mm could achieve RL = –19 dB and EAB = 3.8 GHz [83]. The microwave attenuation mechanisms of the CFO-1/PPy-1.3 absorber could be attributed to the natural resonance frequency, interfacial polarization, dipole polarization, and multiple reflection and scattering.

Overall, bi-component/bi-layer composites of CFO-based materials could improve the microwave absorption performance of CFO. However, the RL values were maintained pretty low, and EAB values were relatively narrow combined with the absorbers' thick thickness, as shown in Figure 4. In order to overcome these issues, CFO-based ternary/quaternary composites should be fabricated, and their microwave absorption properties should be studied, which will be presented in the following sections.

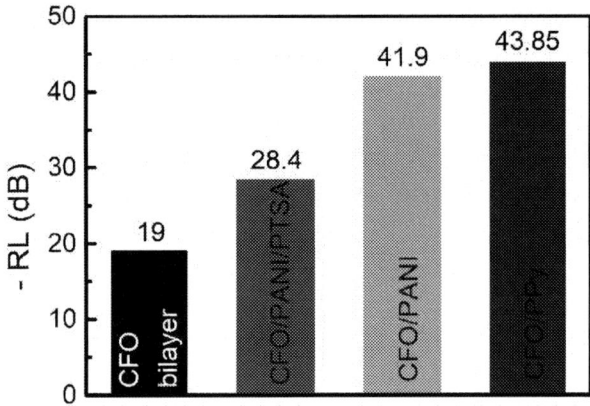

**Figure 4.** Comparison of RL values for CFO/conducting polymer composites.

## Microwave Absorption of CFO-Based Ternary Composites

### Ternary Composite Based on CFO/NFO

As shown in the above section, the CFO/NFO composite could not show excellent microwave absorption performance. Therefore, ternary composites based on CFO/NFO composites were fabricated. Sadeghi et al. reported that the CFO/NFO/carbon fibers (CFs) composite coated polypyrrole significantly enhanced the microwave absorption with an RL of –55 dB at 10.6 GHz and an EAB of 5 GHz [84]. The enhancement in microwave absorption performance of the CFO/NFO/CFs composite could be derived from good impedance matching, resulting from balancing magnetic loss and dielectric loss.

## Ternary Composites Based on CFO and Metals or Alloys

Ge et al. reported that hollow CFO/CoFe@C could achieve RL up to –51 dB and EAB up to 5.9 GHz for $t = 2.17$ mm [85]. A good microwave absorption performance could be achieved due to its hierarchical architecture, which leads to strong attenuation and broad bandwidth. Chen et al. reported that FeCo/CFO/carbon fiber composite could reach an RL of –52.3 dB and an EAB of 5 GHz for $t = 1.95$ mm [86]. Zhang et al. reported that Co-CFO@mesoporous hollow carbon spheres could reach an RL of –65.31 dB and an EAB of 5.76 GHz for $t = 2.1$ mm [87]. The excellent microwave absorption performance could be derived from moderate complex permeability and complex permittivity, leading to excellent impedance matching and multiple loss mechanisms. CoFe-CFO@C could reach an excellent value of RL as –71.73 dB at 4.78 GHz for $t = 3.4$ mm, which could be attributed to its hetero-structure [88].

Su et al. reported that ternary composite of CFO/FeCo/graphite nanosheet could achieve an excellent RL and EAB with RL = –54.3 dB for $t = 1.2$ mm and EAB = 4.08 GHz for $t = 1.3$ GHz [89]. The excellent microwave absorption properties originate from the moderate attenuation capacity and impedance matching.

CoNi alloys could also be used as a component in the CFO composite to enhance microwave absorption. Fang et al. reported that the NiCo-SWCNTs/CFO composite could reach an RL of –47.9 dB at $f = 14.7$ GHz and EAB up to 7.1 GHz (10.5–17.6 GHz) for $t = 1.8$ mm [90]. The enhancement originated from higher permittivity, strong magnetic loss, and good impedance matching. However, when the alloy is FeNi, the composite of FeNi-CFO-PANI showed a pretty poor RL value of –12.2 dB for $t = 2$ mm [91]. Cao et al. reported that a ternary nanocomposite of MWCNTs/CFO/FeCo embedded in PEDOT-PANI co-polymer could turn RL to –90 dB at $f = 13.8$ GHz and EAB to 4 GHz for $t = 1.5$ mm [92]. The enhancements of coupling effects, impedance matching, and interfacial polarization could be used to explain the super microwave absorption performance of MWCNTs/CFO/FeCo. A detailed comparison of the RL and EAB values for different ternary composites based on CFO and metals or alloys is listed in Table 4.

Table 4. Comparison of microwave absorption properties of ternary composites based on CFO and metals or alloys

| Composite | Thickness (mm) | RL (dB) | EAB (GHz) | Refs. |
|---|---|---|---|---|
| CFO/CoFe@C | 2.17 | −51.0 | 5.9 | [85] |
| FeCo/CFO/carbon fiber | 1.95 | −52.3 | 5.0 | [86] |
| Co-CFO@mesoporous hollow carbon | 2.1 | −65.31 | 5.76 | [87] |
| CoFe-CFO@C | 3.4 | −71.73 | 4.78 | [88] |
| CFO/FeCo/graphite | 1.2 | −54.3 | 1.3 | [89] |
| NiCo-SWCNTs/CFO | 1.8 | −47.9 | 7.1 | [90] |
| FeNi-CFO-PANI | 2.0 | −12.2 | - | [91] |
| MWCNTs/CFO/FeCo | 1.5 | −90.0 | 4.0 | [92] |

## Ternary Composites Based on CFO/BaM

Graphene/BaM/CFO could achieve an excellent value of RL as −32.4 dB and EAB of 3 GHz for t = 3 mm [93]. With changes in the third material of composites, PANI composited to CFO/BaM could enhance the RL of −36.4 GHz at 9.8 GHz [94].

## Ternary Composites Based on CFO/$Y_3Fe_5O_{12}$

Garnet ($Y_3Fe_5O_{12}$) could be used as a component in the composite with CFO. However, the composite with two components of CFO and $Y_3Fe_5O_{12}$ could not meet the requirements of an excellent microwave absorbing material. Therefore, the third component could be employed to fabricate ternary microwave absorbing materials. The nanocomposite of graphene/CFO/$Y_3Fe_5O_{12}$ could reach an RL of −36.1 dB at 14.88 GHz and an EAB of 2 GHz (14–16 GHz) for $t$ = 2–3 mm [95]. Another group reported the same microwave absorption properties for graphene/CFO/$Y_3Fe_5O_{12}$ composite with an RL of −32.9 dB at $f$ = 13.6 GHz [96]. If the third component was PANI, the microwave absorption performance was lower than that of graphene with an RL of −23.4 dB at 7.7 GHz [97].

## Ternary Composites Based on CFO/carbon

In order to increase the microwave dissipation energy and decrease the thickness of CFO/CFs-based absorbing materials, Feng et al. prepared a novel hierarchical CFs@CFO@$MnO_2$ using the sol-gel method and hydrothermal reaction. This ternary composite could reach an RL of −34 dB for a thickness of just 1.5 mm, which could be attributed to reasonable electromagnetic matching [98].

Zhao et al. reported that a composite of CFO@C/rGO could achieve a strong RL of −52.5 dB for $t$ = 2 mm and a wide EAB of 5.68 GHz for $t$ = 1.7 mm [99]. The synergetic effects of multiple reflections, magnetic loss, eddy current loss, dielectric loss, natural resonance, interfacial polarization, and charge polarization enhanced microwave dissipation features.

## Ternary Composites Based on CFO/CNTs

Ternary composite aerogel of CFO/CNTs and graphene nanosheets (GNS) could achieve an RL of −29.1 dB at 10.34 GHz for $t$ = 3 mm [100]. The excellent microwave absorption performance could be derived from impedance matching characteristics and eddy current loss. CNTs@CFO/polyimide aerogel could reach an RL value of −54.4 dB and EAB of 7.15 GHz for $t$ = 2.6 mm [101]. The good microwave absorption of this ternary composite could be attributed to its anisotropic structure and the adjustable electromagnetic components.

## Ternary/Quaternary Composites Based On CFO/Graphene

Ternary composite of graphene/PEDOT/CFO could reach RL of −43.2 dB at 9.4 GHz and EAB of 3.1 GHz for t = 2.4 mm [102]. The good microwave dissipation feature could be attributed to interfacial polarization from multi-interfaces, impedance matching, and balancing between magnetic and dielectric losses.

Composite of graphene@CFO@$SiO_2$@$TiO_2$ nanosheets could possess a good RL of −62.8 dB at 6.24 GHz and EAB of 6.24 GHz for $t$ = 4.9 mm [27]. This quaternary composite's excellent microwave absorption performance could have resulted from the triple junctions and multi-interfaces leading to interfacial polarization. In addition, introducing $TiO_2$ to the quaternary composite significantly increased the porosity and specific surface area, leading to more pathways for scattering and reflection of the microwaves.

## Ternary Composites Based on CFO/rGO

Composite of hollow polyacrylonitrile microspheres (PANS), CFO, and rGO could achieve an RL of −14 dB and an EAB of 3.6 GHz (12–15.6 GHz) for $t$ = 2 mm [103]. Ren et al. reported that graphene sheets/rGO-CFO embedded in cyanate ester resin could be reached RL of −21.8 dB at 11.8 GHz and EAB of 2.8 GHz for a thin $t$ of 1.25 mm [104]. The enhancement of microwave absorption properties could be attributed to good impedance matching.

Jaiswal et al. reported that the ternary composite of CFO/rGO/SiO$_2$ could reach an RL of −24.8 dB at 5.8 GHz for $t$ = 2 mm with 60 wt% filler loading [105]. The enhancement of microwave absorption properties could have resulted from the synergistic effects of dielectric and magnetic losses, strong attenuation, and good impedance matching.

Composite of rGO/CFO/ZnS prepared by hydrothermal method combined with co-precipitation method could reach an RL of −43.2 dB for $t$ = 1.8 mm, while the most broadened EAB was 5.5 GHz (10.2–15.7 GHz) for $t$ = 2 mm [106].

The composite of CFO/MWCNTs/rGO could reach an RL value of −46.8 dB at 11.6 GHz for $t$ = 1.6 mm with only 20 wt% filler loading [107]. The good microwave absorption could have resulted from the high surface area, the typical nanostructure with plenty of space, and the synergistic effect.

Zhang et al. reported that introducing lithium-aluminum-silicate (LAS) glass-ceramics to rGO/CFO sheets significantly improved the microwave absorption performance of the composite with an RL of −50 dB for $t$ = 2.3 mm [108]. At the same time, LAS/rGO/CFO with $t$ = 2 mm could achieve the largest EAB value of 6.16 GHz. These excellent values mainly originated from the good impedance matching and polarization of multi-dipole.

Sandwich CFO/rGO/CFO nanocomposites could reach RL of −53.3 dB at 13.92 GHz and EAB of 7.08 GHz for $t$ = 2.6 mm [109]. The excellent microwave dissipation features of the CFO/rGO/CFO composite could have resulted from the synergistic effects of excellent dielectric and magnetic losses and impedance matching.

Composite of rGO-CFO/FeCo NPs showed good microwave absorption frequency of RL as −53.1 dB and EAB as 7 GHz for $t$ = 2.2 mm, which could be attributed to the resonance coupling effect between CFO and FeCo and electric polarization among interfaces of three layers [110].

Wang et al. reported that N-doped rGO aerogels/CFO could reach an excellent RL value of −55.43 dB at 15.36 GHz for $t$ = 2.3 mm and very 10 wt% filler loading [111]. While the largest EAB value of 4.98 GHz (7.76–12.72 GHz) could be achieved for $t$ = 3.3 mm. The enhancement of microwave absorption performance could be derived from specific structure design (beneficial conduction network), polarization relaxation, and impedance matching.

The composite of rGO/CFO/SnS$_2$ could achieve outstanding values of RL as −54.4 dB for $t$ = 1.6 mm and EAB as 4.9 GHz for $t$ = 1.8 mm [112]. This composite's attenuation of microwave energy is attributed to the magnetic loss in the low-range frequency and dielectric loss in the high-range frequency. Composite of rGO/CFO/Ag could reach an RL of −56.5 dB at 13.1 GHz and EAB of 4 GHz (11.1–15.1 GHz) for $t$ = 2.1 mm [113].

Composite of polyvinylpyrrolidone (PVP), rGO, and CFO could reach RL of −56.8 dB at 15.7 GHz for $t$ = 1.96 mm. While the largest EAB of 6.8 GHz (from 10.64 to 17.44 GHz) could be reached for $t$ = 2.2 mm [114]. The good microwave absorption performance could have come from rich interface characteristics, conductive network, dipole polarization, and impedance matching.

Wang et al. reported that 3D porous architecture CFO/N-rGO aerogel could reach RL of −60.4 dB at $f$ = 14.4 GHz for $t$ = 2.1 mm with 20 wt% filler loading [115]. While the largest EAB was 6.48 GHz (11.44–17.92 GHz) with $t$ = 2.2 mm. The excellent microwave absorption properties of CFO/N-rGO originate from its structure with specific surface morphology. Ding et al. reported that the composite of core/shell CFO/PPy and rGO could reach an RL of −50.1 dB at 6.56 GHz with $t$ = 3.6 mm [116]. With the same absorber thickness, CFO/PPy/rGO could achieve a tremendous value of EAB of 13.12 GHz. These excellent RL and EAB values come from impedance matching and attenuation matching.

A one-pot hydrothermal method was employed to synthesize rGO/Sm$_3$Fe$_5$O$_{12}$/CFO ternary nanocomposites, which showed an excellent RL of −73.71 dB at $f$ = 14.88 GHz for $t$ = 2.09 mm [117]. While the thickness of 2.18 mm could reach a significant EAB value of 7.12 GHz (10.8–17.92 GHz). The enhancement in microwave absorption performance could have resulted from the synergistic effect, which could be interfacial and dipole polarization, natural ferromagnetic resonance, eddy current loss, and multiple reflection and scattering.

Ebrahimi-Tazangi et al. reported that $α$-Fe$_2$O$_3$@CFO/GO nanocomposites could reach a super high RL value of −81.24 dB at 11.98 GHz and EAB of 3.78 GHz for $t$ = 1.4 mm. With $t$ = 3.4 mm, this ternary composite could be the largest EAB value of 8.91 GHz and an RL of −75.14 dB [118]. The excellent microwave absorption of this ternary composite was due to synergistic effects of interfacial polarization (from multiple interfaces), multi-reflection and scattering (from the void between the layers), and dielectric loss increase. A detailed comparison of their RL and EAB values is listed in Table 5.

**Table 5.** Comparison of microwave absorption properties of ternary composites based on CFO/rGO

| Composite | Thickness (mm) | RL (dB) | EAB (GHz) | Refs. |
|---|---|---|---|---|
| PANS/CFO/rGO | 2.0 | −14.0 | 3.6 | [103] |
| Graphene sheets/rGO-CFO | 1.25 | −21.8 | 2.8 | [104] |
| CFO/rGO/SiO2 | 2.0 | −24.8 | 5.8 | [105] |
| rGO/CFO/ZnS | 1.8 | −43.2 | 5.5 | [106] |
| CFO/MWCNTs/rGO | 1.6 | −46.8 | - | [107] |
| LAS/rGO/CFO | 2.3 and 2.0 | −50.0 | 6.16 | [108] |
| CFO/rGO/CFO | 2.6 | −53.3 | 7.08 | [109] |
| rGO-CFO/FeCo | 2.2 | −53.1 | 7.0 | [110] |
| N-doped rGO/CFO | 2.3 and 3.3 | −55.43 | 4.98 | [111] |
| rGO/CFO/SnS2 | 1.8 | −54.4 | 4.9 | [112] |
| rGO/CFO/Ag | 2.1 | −56.5 | 4.0 | [113] |
| PVP/rGO/CFO | 1.96 and 2.2 | −56.8 | 6.8 | [114] |
| 3D porous architecture CFO/N-rGO aerogel | 2.1 and 2.2 | −60.4 | 6.48 | [115] |
| CFO/PPy/rGO | 3.6 | −50.1 | 13.12 | [116] |
| rGO/Sm3Fe5O12/CFO | 2.09 and 2.18 | −73.71 | 7.12 | [117] |
| α-Fe2O3@CFO/GO | 1.4 and 3.4 | −81.24 | 8.91 | [118] |

\* For composites with two thickness values: the first number is the thickness for achieving the maximum RL value, while the second is for achieving the maximum EAB value.

**Ternary Composite Based On CFO/Conducting Polymer**

A ternary composite of PPy/CFO/hollow glass microspheres (HGMs) could achieve RL of −14.5 dB for $t$ = 2.58 mm and EAB of 4.02 GHz (8.38–12.4 GHz) [119].

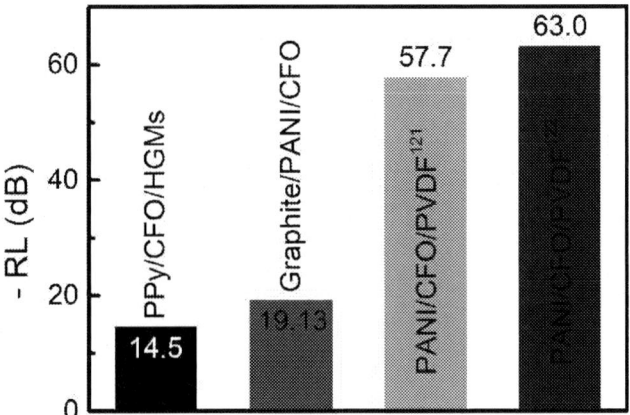

**Figure 5.** Comparison of RL values for ternary composites based on CFO/conducting polymer.

Chen et al. reported that an expanded graphite/PANI/CFO composite could reach a fairly good RL value of −19.13 dB at 13.28 GHz for a fragile 0.5 mm [120]. Yang et al. fabricated a PANI/CFO/polyvinylidene fluoride (PANI/CFO/PVDF) composite using the *in-situ* polarization method, which could reach RL up to −57.7 dB at 7.6 GHz for $t$ = 4 mm and an EAB of 3.4 GHz (6.5–9.9 GHz) [121]. Another group reported similar results with a bit shift in frequency with RL of −63 dB at $f$ = 10.1 GHz and EAB of 2.4 GHz for $t$ = 1.5 mm [122]. The enhancement of microwave absorption performance originated from increased interfacial dipolar polarization and multiple scattering of microwave energy, where PANI/CFO/PVDF could reach the best RL values.

## Conclusion and Future Insights

This book chapter deeply investigated an overview of the microwave absorption enhancement methods of CFO-based materials. Simple methods such as fabrication with different morphologies and doping slightly enhance the microwave absorption performance. Compositing with magnetic materials or conducting polymers also improves microwave absorption performance. Significantly, composites of CFO with allotropies of carbon remarkably increased the microwave absorption properties. In addition, the best microwave absorption performances were achieved in the CFO-based materials' ternary/quaternary composites.

Many methods and materials could still be used to fabricate CFO-based materials to enhance their microwave absorption performance. The microwave absorption properties of CFO-based materials would also be improved for a wide range of practical applications.

## Disclaimer

None

## References

[1] Ma, M., W. Li, Z. Tong, Y. Ma, Y. Bi, Z. Liao, J. Zhou, G. Wu, M. Li, J. Yue, X. Song and X. Zhang, *J. Colloid Interface Sci.* 578, 58-68 (2020).
[2] Cui, Y., K. Yang, J. Wang, T. Shah, Q. Zhang and B. Zhang, Carbon 172, 1-14 (2021).
[3] Di, X., Y. Wang, Y. Fu, X. Wu and P. Wang, *Carbon* 173, 174-184 (2021).
[4] Yan, J., Y. Huang, Y. Yan, X. Zhao and P. Liu, Compos. *Part A Appl. Sci. Manuf.* 139, 106107 (2020).
[5] Zhang, F., Z. Jia, Z. Wang, C. Zhang, B. Wang, B. Xu, X. Liu and G. Wu, Mater. *Today Phys.* 20, 100475 (2021).
[6] Qiu, Y., H. Yang, L. Ma, Y. Lin, H. Zong, B. Wen, X. Bai and M. Wang, *J. Colloid Interface Sci.* 581, 783-793 (2021).
[7] Ni, X., Z. He, X. Liu, Q. Jiao, H. Li, C. Feng and Y. Zhao, *Mater. Lett.* 193, 232-235 (2017).
[8] Shang, T., Q. Lu, L. Chao, Y. Qin, Y. Yun and G. Yun, *Appl. Surf. Sci.* 434, 234-242 (2018).
[9] Tong, Z., Q. Yao, J. Deng, L. Cheng, T. Chuang, J. Wang, G. Rao, H. Zhou and Z. Wang, *Mater. Sci. Eng.* B 268, 115092 (2021).
[10] Lv, H., X. Liang, Y. Cheng, H. Zhang, D. Tang, B. Zhang, G. Ji and Y. Du, *ACS Appl. Mater. Interfaces* 7 (8), 4744-4750 (2015).
[11] Golchinvafa, S., S. M. Masoudpanah and M. Jazirehpour, *J. Alloys Compd.* 809, 151746 (2019).
[12] Zhou, J., X. Shu, Y. Wang, J. Ma, Y. Liu, R. Shu and L. Kong, *J. Magn. Magn. Mater.* 493, 165699 (2020).
[13] Li, W., L. Wang, G. Li and Y. Xu, *J. Magn. Magn. Mater.* 377, 259-266 (2015).
[14] Feng, C., X. Liu, S. W. Or and S. L. Ho, *AIP Adv.* 7 (5), 056403 (2016).
[15] Huang, X., J. Zhang, Z. Liu, T. Sang, B. Song, H. Zhu and C. Wong, *J. Alloys Compd.* 648, 1072-1075 (2015).

[16] Shen, W., B. Ren, S. Wu, W. Wang and X. Zhou, *Appl. Surf. Sci.* 453, 464-476 (2018).
[17] Huang, L., J. Li, Z. Wang, Y. Li, X. He and Y. Yuan, *Carbon* 143, 507-516 (2019).
[18] Li, Y., M. Yuan, H. Liu and G. Sun, *J. Alloys Compd.* 826, 154147 (2020).
[19] Li, W., L. Wang, G. Li and Y. Xu, *J. Alloys Compd.* 633, 11-17 (2015).
[20] Cheng, G. F. Pan, X. Zhu, Y. Dong, L. Cai and W. Lu, *Compos. Commun.* 27, 100867 (2021).
[21] Xu, R., D. Xu, Z. Zeng and D. Liu, *Chem. Eng. J.* 427, 130796 (2022).
[22] Feng, J. Y. Wang, Y. Hou, J. Li and L. Li, *Ceram. Int.* 42 (15), 17814-17821 (2016).
[23] Li, G., L. Sheng, L. Yu, K. An, W. Ren and X. Zhao, *Mater. Sci. Eng. B* 193, 153-159 (2015).
[24] Wang, S., Y. Zhao, H. Xue, J. Xie, C. Feng, H. Li, S. Shi, S. Muhammad and Q. Jiao, *Mater. Lett.* 223, 186-189 (2018).
[25] Zhang, S., Q. Jiao, J. Hu, J. Li, Y. Zhao, H. Li and Q. Wu, *J. Alloys Compd.* 630, 195-201 (2015).
[26] Zhang, M., J. Zhang, H. Lin, T. Wang, S. Ding, Z. Li, J. Wang, A. Meng, Q. Li and Y. Lin, *Compos. B. Eng.* 190, 107902 (2020).
[27] Zong, M., Y. Huang and N. Zhang, *Appl. Surf. Sci.* 345, 272-278 (2015).
[28] Liu, Y., Z. Chen, Y. Zhang, R. Feng, X. Chen, C. Xiong and L. Dong, *ACS Appl. Mater. Interfaces* 10 (16), 13860-13868 (2018).
[29] Zhang, K., J. Li, F. Wu, M. Sun, Y. Xia and A. Xie, *ACS Appl. Nano Mater.* 2 (1), 315-324 (2019).
[30] He, J., S. Liu, L. Deng, D. Shan, C. Cao, H. Luo and S. Yan, *Appl. Surf. Sci.* 504, 144210 (2020).
[31] Zhao, X., Y. Huang, X. Liu, J. Yan, L. Ding, M. Zong, P. Liu and T. Li, *J. Colloid Interface Sci.* (2021).
[32] Su, X., J. Wang, X. Zhang, S. Huo, W. Chen, W. Dai and B. Zhang, *Ceram. Int.* 46 (8, Part B), 12353-12363 (2020).
[33] Chen, J., J. Zheng, Q. Huang, F. Wang and G. Ji, *ACS Appl. Mater. Interfaces* 13 (30), 36182-36189 (2021).
[34] Ge, J., S. Liu, L. Liu, Y. Cui, F. Meng, Y. Li, X. Zhang and F. Wang, *J. Mater. Sci. Technol.* 81, 190-202 (2021).
[35] Zhang, Y. M. Yao, C. Liu, H. Zhao, X. Miao and F. Xu, *ACS Appl. Nano Mater.* 3 (9), 8939-8948 (2020).
[36] Ni, X., Z. He, X. Liu, Q. Jiao, H. Li, C. Feng and Y. Zhao, *Mater. Lett.* 193, 232-235 (2017).
[37] Shang, T., Q. Lu, L. Chao, Y. Qin, Y. Yun and G. Yun, *Appl. Surf. Sci.* 434, 234-242 (2018).
[38] Jiang, P., Q. Xu, N. Tran, A. S. El-Shafay, V. Mohanavel, A. Abdelrahman and M. Ravichandran, *Ceram. Int.* 48 (10), 13541-13550 (2022).
[39] Jafarian, M., S. F. Kashani-Bozorg, A. A. Amadeh and Y. Atassi, *Ceram. Int.* 47 (21), 30448-30458 (2021).
[40] Tong, Z., Q. Yao, J. Deng, L. Cheng, T. Chuang, J. Wang, G. Rao, H. Zhou and Z. Wang, *Mater. Sci. Eng. B* 268, 115092 (2021).
[41] Feng, C., X. Liu, S. W. Or and S. L. Ho, *AIP Adv.* 7 (5), 056403 (2016).

[42]  Mahdikhah, V., A. Ataie, A. Babaei, S. Sheibani, C. W. Ow-Yang and S. K. Abkenar, *Ceram. Int.* 46 (11, Part A), 17903-17916 (2020).
[43]  Golchinvafa, S., S. M. Masoudpanah and M. Jazirehpour, *J. Alloys Compd.* 809, 151746 (2019).
[44]  Liu, Y., Y. Su, J. Zhang, Q. Dong and C. Shi, *J. Supercond. Nov. Magn.* 34 (8), 2217-2225 (2021).
[45]  Zhou, J., X. Shu, Y. Wang, J. Ma, Y. Liu, R. Shu and L. Kong, *J. Magn. Magn. Mater.* 493, 165699 (2020).
[46]  Huang, X., J. Zhang, Z. Liu, T. Sang, B. Song, H. Zhu and C. Wong, *J. Alloys Compd.* 648, 1072-1075 (2015).
[47]  Zhang L. and Z. Li, *J. Alloys Compd.* 469 (1), 422-426 (2009).
[48]  Tyagi, S., H. B. Baskey, R. C. Agarwala, V. Agarwala and T. C. Shami, *Trans. Indian Inst. Met.* 64 (3), 271-277 (2011).
[49]  Azizi, P., S. M. Masoudpanah and S. Alamolhoda, *Appl. Phys. A* 125 (10), 686 (2019).
[50]  Tursun, R., Y. Su, J. Zhang and R. Yakefu, *J. Alloys Compd.* 911, 165051 (2022).
[51]  Lv, H., X. Liang, Y. Cheng, H. Zhang, D. Tang, B. Zhang, G. Ji and Y. Du, *ACS Appl. Mater. Interfaces* 7 (8), 4744-4750 (2015).
[52]  Salimkhani, H., A. Motei Dizaji, E. Hashemi, P. Palmeh, G. Sabeghi and S. Salimkhani, *Ceram. Int.* 42 (11), 12709-12714 (2016).
[53]  Zhou, J., R. Tan, Z. Yao, H. Lin and Z. Li, *Mater. Chem. Phys.* 244, 122697 (2020).
[54]  Gan, W., L. Gao, W. Zhang, J. Li and X. Zhan, *Ceram. Int.* 42 (11), 13199-13206 (2016).
[55]  Li, Y., M. Yuan, H. Liu and G. Sun, *J. Alloys Compd.* 826, 154147 (2020).
[56]  Liu, Z., C. Shi, F. He and N. Zhao, *Integr. Ferroelectr.* 208 (1), 164-176 (2020).
[57]  Xu, J., D. Liu, Y. Meng, S. Tang, F. Wang, C. Bian, X. Chen, S. Xiao, X. Meng and N. Yang, *Nanotechnology* 33 (21), 215603 (2022).
[58]  Xu, R., D. Xu, Z. Zeng and D. Liu, *Chem. Eng. J.* 427, 130796 (2022).
[59]  Cheng, G., F. Pan, X. Zhu, Y. Dong, L. Cai and W. Lu, *Compos. Commun.* 27, 100867 (2021).
[60]  Feng, J., Y. Wang, Y. Hou, J. Li and L. Li, *Ceram. Int.* 42 (15), 17814-17821 (2016).
[61]  Zhang, B. B., P. F. Wang, J. C. Xu, Y. B. Han, H. X. Jin, D. F. Jin, X. L. Peng, B. Hong, J. Li, J. Gong, H. L. Ge, Z. W. Zhu and X. Q. Wang, *Nano* 10 (05), 1550070 (2015).
[62]  Li, G., L. Sheng, L. Yu, K. An, W. Ren and X. Zhao, *Mater. Sci. Eng. B* 193, 153-159 (2015).
[63]  Yuan, Y., S. Wei, Y. Liang, B. Wang, Y. Wang, W. Xin, X. Wang and Y. Zhang, *J. Alloys Compd.* 867, 159040 (2021).
[64]  Zhang, S., Z. Qi, Y. Zhao, Q. Jiao, X. Ni, Y. Wang, Y. Chang and C. Ding, *J. Alloys Compd.* 694, 309-312 (2017).
[65]  Liu, G., L. Wang, H. Zhang, Z. Du, X. Zhou, K. Wang, Y. Sun and S. Gao, *Compos. Commun.* 27, 100910 (2021).
[66]  Ashfaq, M. Z., A. Ashfaq, M. K. Majeed, A. Saleem, S. Wang, M. Ahmad, M. M. Hussain, Y. Zhang and H. Gong, *Ceram. Int.* 48 (7), 9569-9578 (2022).

[67] Yadav, P., S. Rattan, A. Tripathi and S. Kumar, *Ceram. Int.* 46 (1), 317-324 (2020).
[68] Fu, M., Q. Jiao, Y. Zhao and H. Li, *J. Mater. Chem.* A 2 (3), 735-744 (2014).
[69] Li, X., J. Feng, H. Zhu, C. Qu, J. Bai and X. Zheng, *RSC Adv.* 4 (63), 33619-33625 (2014).
[70] Zhang, S., Q. Jiao, J. Hu, J. Li, Y. Zhao, H. Li and Q. Wu, *J. Alloys Compd.* 630, 195-201 (2015).
[71] Ma, T., Y. Cui, L. Liu, H. Luan, J. Ge, P. Ju, F. Meng and F. Wang, *RSC Adv.* 10 (53), 31848-31855 (2020).
[72] Guan, X. H., J. M. Kuang, L. Yang, M. Lu and G. S. Wang, *ChemistrySelect* 4 (33), 9516-9522 (2019).
[73] Moitra, D., M. Chandel, B. K. Ghosh, R. K. Jani, M. K. Patra, S. R. Vadera and N. N. Ghosh, *RSC Adv.* 6 (80), 76759-76772 (2016).
[74] Liu, Z., G. Xu, M. Zhang, K. Xiong and P. Meng, *J. Mater. Sci.: Mater. Electron.* 27 (9), 9278-9285 (2016).
[75] Liu, Y., Q. Wang, Q. Fang, W. Liu and F. Meng, *J. Magn. Magn. Mater.* 546, 168903 (2022).
[76] Zong, M., Y. Huang, H. Wu, Y. Zhao, Q. Wang and X. Sun, *Mater. Lett.* 114, 52-55 (2014).
[77] Zong, H., H. Yang, J. Dong, L. Ma, Y. Lin, Y. Qiu and B. Wen, *J. Mater. Sci.: Mater. Electron.* 31 (21), 18590-18604 (2020).
[78] Zhang, M., J. Zhang, H. Lin, T. Wang, S. Ding, Z. Li, J. Wang, A. Meng, Q. Li and Y. Lin, *Compos. B. Eng.* 190, 107902 (2020).
[79] He, J., S. Liu, L. Deng, D. Shan, C. Cao, H. Luo and S. Yan, *Appl. Surf. Sci.* 504, 144210 (2020).
[80] Ismail, M. M., S. N. Rafeeq, J. M. A. Sulaiman and A. Mandal, *Appl. Phys. A* 124 (5), 380 (2018).
[81] Praveena K. and M. Bououdina, *J. Electron. Mater.* 49 (10), 6187-6198 (2020).
[82] Ren, X., J. Wang, H. Yin, Y. Tang, H. Fan, H. Yuan, S. Cui and L. Huang, *Appl. Surf. Sci.* 575, 151752 (2022).
[83] Zhang, L., N. Stalin, N. Tran, S. Mehrez, M. F. Badran, V. Mohanavel and Q. Xu, *Ceram. Int.* 48 (11), 16374-16385 (2022).
[84] Sadeghi, R., A. Sharifi, M. Orlowska and I. Huynen, *Micromachines* 11 (9) (2020).
[85] Ge, J., S. Liu, L. Liu, Y. Cui, F. Meng, Y. Li, X. Zhang and F. Wang, *J. Mater. Sci. Technol.* 81, 190-202 (2021).
[86] Chen, J., J. Zheng, Q. Huang, F. Wang and G. Ji, *ACS Appl. Mater. Interfaces* 13 (30), 36182-36189 (2021).
[87] Zhang, H., Z. Jia, B. Wang, X. Wu, T. Sun, X. Liu, L. Bi and G. Wu, *Chem. Eng. J.* 421, 129960 (2021).
[88] Guan, Z. J., J. T. Jiang, N. Chen, Y. X. Gong and L. Zhen, *Nanotechnology* 29 (30), 305604 (2018).
[89] Su, X., J. Wang, X. Zhang, S. Huo, W. Chen, W. Dai and B. Zhang, *Ceram. Int.* 46 (8, Part B), 12353-12363 (2020).
[90] Fang, Y. H., X. T. Tang, X. Sun, Y. F. Zhang, J. W. Zhao, L. M. Yu, Y. Liu and X. L. Zhao, *J. Appl. Phys.* 121 (22), 224301 (2017).

[91]  Li, Z., J. Wang and F. Zhao, *IOP Conf. Ser.: Mater. Sci. Eng.* 678 (1), 012143 (2019).
[92]  Cao, Y., N. Farouk, N. Mortezaei, A. V. Yumashev, M. N. Akhtar and A. Arabmarkadeh, *Ceram. Int.* 47 (9), 12244-12251 (2021).
[93]  Yang, H., T. Ye, Y. Lin and M. Liu, *Appl. Surf. Sci.* 357, 1289-1293 (2015).
[94]  Yang, H., T. Ye, Y. Lin and M. Liu, *J. Alloys Compd.* 653, 135-139 (2015).
[95]  Wang, Y., Y. Pu, Y. Shi and H. Cui, *J. Mater. Sci.: Mater. Electron.* 28 (17), 12866-12872 (2017).
[96]  Lin, Y., X. Liu, T. Ye, H. Yang, F. Wang and C. Liu, *J. Mater. Sci.: Mater. Electron.* 27 (8), 8177-8182 (2016).
[97]  Lin, Y., X. Liu, T. Ye, H. Yang, F. Wang and C. Liu, *J. Mater. Sci.: Mater. Electron.* 27 (5), 4833-4838 (2016).
[98]  Feng, A., T. Hou, Z. Jia and G. Wu, *RSC Adv.* 10 (18), 10510-10518 (2020).
[99]  Zhao, X., Y. Huang, X. Liu, J. Yan, L. Ding, M. Zong, P. Liu and T. Li, *J. Colloid Interface Sci.* 607, 192-202 (2022).
[100] Ren, F., Z. Guo, Y. Shi, L. Jia, Y. Qing, P. Ren and D. Yan, *J. Alloys Compd.* 768, 6-14 (2018).
[101] Luo, J., Y. Wang, Z. Qu, W. Wang and D. Yu, *Chem. Eng. J.* 442, 136388 (2022).
[102] Liu, P., Y. Huang and X. Zhang, *Powder Technol.* 276, 112-117 (2015).
[103] Zhang, B., J. Wang, J. Wang, H. Duan, S. Huo and Y. Tang, *J. Mater. Sci.: Mater. Electron.* 28 (4), 3337-3348 (2017).
[104] Ren, F., G. Zhu, P. Ren, K. Wang, X. Cui and X. Yan, *Appl. Surf. Sci.* 351, 40-47 (2015).
[105] Jaiswal, R., K. Agarwal, V. Pratap, A. Soni, S. Kumar, K. Mukhopadhyay and N. Eswara Prasad, *Mater. Sci. Eng. B* 262, 114711 (2020).
[106] Zhang, N., Y. Huang, M. Zong, X. Ding, S. Li and M. Wang, *Chem. Eng. J.* 308, 214-221 (2017).
[107] Zhang, K., X. Gao, Q. Zhang, T. Li, H. Chen and X. Chen, *J. Alloys Compd.* 723, 912-921 (2017).
[108] Zhang, X., L. Xia, B. Zhong, H. Yang, B. Shi, L. Huang, Y. Yang and X. Huang, *J. Alloys Compd.* 799, 368-376 (2019).
[109] Zhang, K., J. Li, F. Wu, M. Sun, Y. Xia and A. Xie, *ACS Appl. Nano Mater.* 2 (1), 315-324 (2019).
[110] Zhang, Y., M. Yao, C. Liu, H. Zhao, X. Miao and F. Xu, *ACS Appl. Nano Mater.* 3 (9), 8939-8948 (2020).
[111] Wang, X., J. Liao, R. Du, G. Wang, N. Tsidaeva and W. Wang, *J. Colloid Interface Sci.* 590, 186-198 (2021).
[112] Zhang, N., Y. Huang, M. Zong, X. Ding, S. Li and M. Wang, *Ceram. Int.* 42 (14), 15701-15708 (2016).
[113] Huang, Y., X. Zhao, M. Zong, J. Yan and T. Li, *J. Mater. Sci. Mater. Electron.* (2021).
[114] Zhu, T., S. Chang, Y.-F. Song, M. Lahoubi and W. Wang, *Chem. Eng. J.* 373, 755-766 (2019).
[115] Wang, X., Y. Lu, T. Zhu, S. Chang and W. Wang, *Chem. Eng. J.* 388, 124317 (2020).

[116] Ding, L., X. Zhao, Y. Huang, J. Yan, T. Li and P. Liu, *J. Colloid Interface Sci.* 595, 168-177 (2021).
[117] Shen, W., B. Ren, S. Wu, W. Wang and X. Zhou, *Appl. Surf. Sci.* 453, 464-476 (2018).
[118] Ebrahimi-Tazangi, F., J. Seyed-Yazdi and S. H. Hekmatara, *J. Alloys Compd.* 900, 163340 (2022).
[119] Wang, X., H. Yan, R. Xue and S. Qi, *J. Mater. Sci.: Mater. Electron.* 28 (1), 519-525 (2017).
[120] Chen, K., C. Xiang, L. Li, H. Qian, Q. Xiao and F. Xu, *J. Mater. Chem.* 22 (13), 6449-6455 (2012).
[121] Yang, H., N. Han, Y. Lin, G. Zhang and L. Wang, *RSC Adv.* 6 (102), 100585-100589 (2016).
[122] Dabas, S., M. Chahar and O. P. Thakur, *Mater. Chem. Phys.* 278, 125579 (2022).

## Chapter 2

# Strong Resonance Effects in Ordered Layered Photonic Structures for Filtering, Collimation, Metrology and Spectroscopy

### E. Ya. Glushko[*]
Institute of Semiconductor Physics of NAS of Ukraine, Kyiv, Ukraine

## Abstract

A planar photonic resonator containing unitary defect in the middle of the structure can exhibit a system of extraordinary narrow resonance peaks of transmission on the background of perfect reflection. The properties of standing modes inside the polyethylene (polypropylene)/ silicon plane resonators in the total intrinsic reflection region and unusual manifestations of THz transmission spectra in centimeter and millimeter wavelength range were studied. It is shown that the angle and frequency half-widths of the resonance peak can be less than $10^{-9}$ of the magnitude of angle and frequency in dependence on the number of periods. This allows one to form collimated beams with the divergence measured in a fraction of a microdegree. It is shown that a plane resonator containing a central defect transforms the frequency divided peaks into the outgoing-transmitted beams of various directions like a prism transforms light. This opens the way for precision measurements of angle and frequency distribution of THz radiation. It is proposed to use the existing extremely sharp peaks of transmission in planar resonators containing a central defect for aims of spectroscopy and metrology. A new spectroscopy technique is proposed based on the existing sharp transmission resonances using the conception of accumulating reservoir of

---

[*] Corresponding Author's Email: scientist.com_eugene.glushko@mail.com.

In: Electromagnetic Waves
Editor: Manuel B. Hutchinson
ISBN: 979-8-88697-254-2
© 2022 Nova Science Publishers, Inc.

electromagnetic field. Extra-high extent of collimation resulting from the usage of defected photonic resonators gives an opportunity to form long and stable channels of communication in the THz frequency range.

**Keywords:** photonic crystal resonator, THz waves, THz spectroscopy, layered structures, collimation effect

## Introduction

Photonic structures have a wide spectrum of applications in science and technology such as night-vision applications, safety and defense, sensing, communication applications, signal processing and many other areas [1–8]. A great attention attracts the resonator properties of photonic structures. Some extraordinary forms of reflection and transmission occur inside the area of perfect reflective reflection windows occur in photonic crystal (PhCr) resonators accompanied by extremely sharp local resonances of transmission [9-12]. It was found that if a polarized electromagnetic wave (EMW) falls onto a PhCr, the presence of metal there generates a system of narrow spectral holes inside the reflection windows existing for the same metal-free photonic crystal. An investigation of properties including the resonances shape showed that they are of Fano type at small angles of incidence independently from polarization. Quite another manifestation of metal-resonator inter-influence takes place at whispering incident angles when reflection spikes of a p-polarized wave coincide with the modes of the photonic crystal resonator and they are absent throughout the stop-band areas between modes. That kind of behavior is typical for a surface plasmon (SP) resonance. The s-polarized field exhibited more complicate features: at whispering incident angles the resonances in vicinity of low energy modes of each band arise between them (Fano type resonances) whereas the resonances with higher energy close to the top of band became to be matched the mode frequencies (SP resonances).

The 1D photonic crystals (PhCr) in photonics can play to a certain extent a role similar to that of hydrogen atom in quantum mechanics of atomic spectra: they have a simple analytical solution of the eigenstates problem and can also be used to classify the photonic states for the more complex 2D and 3D photonic structures. An attempt to build a classification scheme for a two-dimensional photonic resonator was undertaken in [13] on the ground of an exact basis of eigenfunctions for a model 1D structure. A general description of reflection spectra for metalized photonic resonators was undertaken in [12,

14] where an effect of needle-like dips of reflection – piercing - accompanied by enhanced resonant absorption of electromagnetic energy in a metal film was found. It was shown there that the arising sharp dips in reflection have interference nature: they originate due to a property of resonator to concentrate electromagnetic energy immediately near the metalized end of the resonator.

In general, the sharp resonant dips of reflection against the background of ideally reflecting spectral areas in metalized PhCr as well as the resonant peaks of transmission in the defected resonators [15] are by its nature the interference induced phenomena in periodic structures: the sharpness of peaks or spikes increases with growth of the number of periods. The 1D PhCr resonators are also convenient to study the practical applications of local states, both surface and intrinsic ones, existing in defected resonators. Very attractive are their technological simplicity and an opportunity to forecast the parameters of electromagnetic devices. Due to the well known spectral scaling property of the 1D periodic structures [16], one can fabricate resonators with needed spectrum and tune the spectrum structure including the corresponding local states. The use of latter instead of the described in [17] shifting band method can give an essential advantage in energy of a signal needed to perform logic operations. The reason is that local states of a resonator with doubled defect may be situated close enough one to another and can give a sharp transfer from perfect reflection to perfect transmission. An important circumstance is also an opportunity to create the optically high contrastive structures with practically absent losses of electromagnetic energy in this frequency region [18]. One more important technological aspect is that due to the scaling rule acting for the ordered 1D photonic structures there is no difficulty to calculate the sizes of a resonator for the needed frequency interval. In general, the existence of a local electromagnetic state is a manifestation of high optical contrastivity of the structure. In the case of surface states it is a contrastivity regarding external medium. The intrinsic local state erasing inside a defect containing resonator indicates a high optical contrast between defect and photonic resonator. It worth noting that though the optics of 1D photonic structures was analyzed in a lot of publications nevertheless the THz frequency region cannot be considered as a comprehensive studied area. First, it concerns a physical relationship between the modes of total intrinsic reflection (TIR) region and spectra of reflection measured in external geometry of incidence out the TIR region. Besides, it touches the physical nature of sharp absorption resonances of polarized EMW piercing a metal film contacting the resonator [12, 14]. The peaks of resonant transmission against the backdrop of the perfect reflection windows in visual and IR frequency region were considered in [10, 11, 19, 20] for some partial cases at the normal incidence only and regardless the mode spectrum of resonator. A more comprehensive study of phenomena accompanying the EMW behavior in defected/metalized resonators for both external reflection spectra and intrinsic resonator modes structure in a more wide frequency region was presented in [12-16] to recognize the entire operational field more exactly. This very approach has given an opportunity to consider some new effects of filtering and collimating. Note that the resonator modes, i.e., standing waves captured inside the total internal region (TIR) of the photonic resonator, cannot be considered as well-researched in literature. In addition to some inertia of the research frontline, a reason may lie in some experimental difficulties of excitation and observation of these states inside the TIR region.

An important component of investigation scheme of resonance phenomena is the use of so called electromagnetic box – a lossless reservoir of electromagnetic field (EMR) – that serves to accumulate the radiation up to the state of saturation with the uniform angular distribution. This stochastic EMW reservoir can accumulate inside the electromagnetic energy from a source of intensity $J_l$ and the fixed frequency $\upsilon$. We suppose that due to stochastic inner reflection the angle distribution of irradiation inside the reservoir is of the white noise type. When the average density of energy inside the box reaches its saturation a stable energy flow $J_r$ through the metalized resonator window is established. The outgoing flow outside the EMR is determined by the accumulated density of electromagnetic energy W inside the reservoir. In the considered millimeter range of wavelength, we will neglect the absorption of energy by the EMR walls. Therefore, one can take into account only two channels of energy leakage: through the side faces of the PhCr and by the reverse flow losses through the source window.

As to the terminology, we will call local modes - the resonator eigenstates which can exist only inside the TIR region whereas the narrow peaks of transmission in geometry of incidence shown in Figure 1 should be called resonances [14].

In this study, we consider different resonant effects in planar central defect containing photonic crystal resonators that are able to excite both standing waves inside TIR region and waves transmitting through and reflected from the structure. It was found that a system of extraordinary narrow resonance peaks of transmission on the background of perfect reflection can be observed in photonic resonators of such kind. The properties of both standing modes inside the polyethylene (polypropylene)/silicon plane resonators region and unusual manifestations of THz transmission spectra in centimeter and millimeter wavelength range will be studied. A way of spectroscopy based on the existing sharp transmission resonances using the concept of accumulating reservoir of electromagnetic field will be analyzed. Potential applications of the found strong resonant effects in the THz spectroscopy, collimating and filtering devices, metrology, and communication lines are discussed.

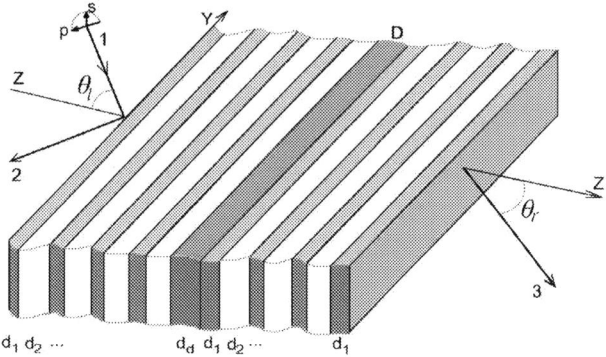

**Figure 1**. A layered photonic resonator in YZ plane. External incidence. A 7 period binary structure. An asymmetric defect layer D of thickness $d_d$ occupies asymmetric position $N_d$=9, $d_1$, $d_2$, layer thicknesses, the PhCr period $d_1+ d_2$. Angle of incidence $\theta_l$, output angle $\theta_r$. 1, incident beam: s(p), electric field vector in a s(p)-polarized wave, 2, reflected beam, 3, transmitted beam. YX plane is considered as infinite.

## Defect Containing Photonic Resonator: Bandgap Structure and Map of Reflection

We analyze here the p- and s-polarized EMW in a layered N-period silicon-polyethylene (or silicon-polypropylene) Si/PE(PP) structure containing an intrinsic central defect layer. The magnitude of thicknesses for Si layers $d_1$ and PP (or PE) voids $d_2$ were chosen to consider the bandgap structure and effects of resonant transmission in the THz frequency region. The position of a defect layer of thickness $d_D$ corresponds to number $2N_d+1$. It means that a polarized EMW passes through $N_d$ periods of the PhCr, meets the defect layer, and then propagates through the remaining $N-N_d$ periods of the structure with an additional ending Si-layer of same size (Figure 1).

According to the chosen geometry, a p- or s-polarized plane electromagnetic wave with the refraction vector $\bm{n_s}=n_s \ (cos\theta_s, \ sin\theta_s)$ inside an arbitrary s-layer can be expressed in a view containing Z and Y components:

$$e^{-ik_{sy}z}k_s\mathbf{E}_s = (-k_{sy}, k_{sz})A_s e^{ik_{sz}z} - (k_{sy}, k_{sz})B_s e^{-ik_{sz}z}, \qquad (1)$$

where $A_s$, $B_s$ are electric field amplitudes, z indicates projection of refraction vector and wave vector $\mathbf{k}_s$ into axes OZ in geometry of p-polarization. The light passing through the system is determined by Maxwell boundary conditions of continuous tangential components of fields is obeyed a self-consistent chain of 2N+6 matrix equations for field amplitudes A and B (from left to right):

$$\begin{cases} \mathbf{M}_l \begin{pmatrix} A_l \\ B_l \end{pmatrix} = \mathbf{L}_1 \begin{pmatrix} A_1 \\ B_1 \end{pmatrix}; \quad \mathbf{M}_1 \begin{pmatrix} A_1 \\ B_1 \end{pmatrix} = \mathbf{L}_2 \begin{pmatrix} A_2 \\ B_2 \end{pmatrix}; \\ \mathbf{M}_2 \begin{pmatrix} A_2 \\ B_2 \end{pmatrix} = \mathbf{L}_3 \begin{pmatrix} A_3 \\ B_3 \end{pmatrix}; \quad \mathbf{M}_2 \begin{pmatrix} A_2 \\ B_2 \end{pmatrix} = \mathbf{L}_3 \begin{pmatrix} A_3 \\ B_3 \end{pmatrix}; \\ \cdots \\ \mathbf{M}_{2N_d} \begin{pmatrix} A_{2N_d} \\ B_{2N_d} \end{pmatrix} = \mathbf{L}_{2N_d+1} \begin{pmatrix} A_{2N_d+1} \\ B_{2N_d+1} \end{pmatrix}; \quad \mathbf{M}_{2N_d+1} \begin{pmatrix} A_{2N_d+1} \\ B_{2N_d+1} \end{pmatrix} = \mathbf{L}_{2N_d+2} \begin{pmatrix} A_{2N_d+2} \\ B_{2N_d+2} \end{pmatrix}; \\ \cdots \\ \mathbf{M}_{2N+4} \begin{pmatrix} A_{2N+1} \\ B_{2N+1} \end{pmatrix} = \mathbf{L}_{2N+2} \begin{pmatrix} A_{2N+2} \\ B_{2N+2} \end{pmatrix}; \quad \mathbf{M}_{2N+2} \begin{pmatrix} A_{2N+2} \\ B_{2N+2} \end{pmatrix} = \mathbf{L}_r \begin{pmatrix} A_r \\ B_r \end{pmatrix} \end{cases} \qquad (2)$$

where indexes $l$ and $r$ mark the left and right media, correspondingly; $A_r$ is the amplitude of the wave running away into the right-hand semi-space at $\theta_r$ output angle whereas the amplitude $B_r$ should be set equal to zero. The central row in (2) describes the amplitudes $2N_d+1$ and $2N_d+2$ corresponding to a defect layer inside the structure. The system of equations (2) is written for the polarized field and includes all possible cases of passing through and reflected from the structure electromagnetic waves. If the electromagnetic field is excited inside the total internal region of the PhCr through the input prism (internal problem) we must take $A_l=0$ and equation (2) is transformed into the eigenvalue problem describing spectrum of standing waves and their density distribution in space - eigenfunctions. If the beam 1 falls onto the resonator like it is shown in Figure 1, the module of amplitude $A_l$ should be taken equal to *1* and therefore all other amplitudes have an absolute meaning. In this case, the system (2) issues a continuous spectrum and can be described in terms of reflection and transmission (external problem). The entrance phase can also be taken into account.

The matrices of transfer through the right boundary $\mathbf{M}_s$ and left boundary $\mathbf{L}_s$ of the s-layer are:

$$\mathbf{M}_s = \begin{pmatrix} \cos\theta_s e^{ik_{sz}d_s}, & -\cos\theta_s e^{-ik_{sz}d_s} \\ k_{sz}e^{ik_{sz}d_s}, & k_{sz}e^{-ik_{sz}d_s} \end{pmatrix}, \quad \mathbf{L}_s = \begin{pmatrix} \cos\theta_s, & -\cos\theta_s \\ k_{sz}, & k_{sz} \end{pmatrix} \quad (3a)$$

for p-polarized waves and

$$\tilde{\mathbf{M}}_s = \begin{pmatrix} e^{ik_{sz}d_s}, & -e^{-ik_{sz}d_s} \\ k_{sz}e^{ik_{sz}d_s}, & k_{sz}e^{-ik_{sz}d_s} \end{pmatrix}, \quad \tilde{\mathbf{L}}_s = \begin{pmatrix} 1, & -1 \\ k_{sz}, & k_{sz} \end{pmatrix} \quad (3b)$$

for s-polarized waves, where $d_s$ is the width of an arbitrary s-layer.

We have considered the different defect containing structures to study the phenomenon of sharp resonant transmission of polarized THz radiation arising inside the windows of perfect reflection. In the TIR angular region, the spectrum consists of bands of eigenmodes separated by energy gaps. These modes physically are standing waves and they form a complete system of eigenfunctions describing field distribution inside the resonator depending on the number of photonic eigenstate and incident angle. The completeness property means that orthogonality and node theorem for eigenstates of any nature are valid. In the simplest case, the first band includes states with number of nodes from 0 (lowest state) to N-1. The second band, correspondingly, contains states with number of nodes from N (lowest state) to 2N-1. Every next state is of higher energy at given angle and has one node more in the field distribution along the resonator. In a more general case, the number of modes in a band may be not equal to N in general because the surface and defect local modes can to detach at the appropriate angles of incidence from area band into the gap area. It is worth noting that excitation of resonator's electromagnetic eigenstates needs some special ways (prism, not shown in Figure 1) of introducing the radiation inside the TIR region. If the surrounding medium is air, the critical angles of total internal reflection are approximately 17.58°, 50.66°, 54.59° for field in silicon, polyethylene and polypropylene, correspondingly.

Another experimental case corresponds to geometry of external incidence (Figure 1) when all incident angles that are outside the TIR region and field inside the resonator has a character of not standing but transmitted waves. The physics in this case is described in terms of transmission and reflection in absence of losses. The general view of reflection in geometry of external incidence – reflection map – is genetically linked with the bandgap structure of the TIR region and consists of well expressed windows of reflection and areas of transmission. The presence of a central defect is manifested on the reflection map as a system of narrow resonant lines of transmission against the background of reflection windows. In the case of non-central position of the defect, the resonance becomes weak: the maximum of transmission decreases and its half width at half maximum (HWHM) increases.

To show more distinctively the spectrum and reflection properties we present in Figure 2 a resonator with small number of periods. In figures 2a, 2b, the calculated by system (2) spectrum of the 10 period $(Si/PE)_5/D$ $(Si/PE)_5/Si$ photonic resonator is shown for p- and s-polarized field both inside the TIR region $\theta_1 \in (17.58°, 90°)$ (top panel, <37.58°) and for external incidence $\theta_1 \in (0°, 90°)$ (lower panel). The resonator under consideration contains a central silicon defect D of thickness 40 µm and has period 110 µm with polyethylene voids size 70 µm and silicon layers width 40 µm. The chosen frequency interval (0, 1000) GHz contains two bands: first of them with 10 - 11 resonator band eigenstates and from 1 to 3 detached from both bands local states. The wider interval (0, 3) THz includes four bands of photonic states divided by gaps. The visible paired structure of modes inside the band is a result of resonator structure: a defect cuts PhCr onto two equivalent parts having the same spectrum. In Figure 2a, top panel, presented are the bandgap structure of p-polarized field inside a 10-period defected Si/PE photonic crystal. A complicate behavior of different types of eigenmodes can be observed: band states exhibit paired structure, local states (surface and defect) detach from or confluence into the bands. One more transformation is realized at $\theta_1 > 28.30°$ when the TIR condition at Si/PE boundary is fulfilled: sinusoidal standing waves inside PE layers become hyperbolical (horizontal arrow). We do pay so much attention to the TIR region states because a tight link between the TIR and open divisions of the angle space exists. In particular, all windows of perfect reflection in geometry of external incidence are to some extent a continuation of energy gaps in the region of TIR (upper panel/lower panel) and local eigenstates of the resonator continue their existence outside the TIR as resonances. The local states of two types exist in the structure. Our calculations show that the p-polarized local states begin to detach from the band bottom beginning with $d_D$ approximately 15% of $d_{Si}$. A thick defect layer D ($d_D > 3.5 d_{Si}$) forms 4 local states detaching in series at bigger angles from the bottom of each band. For instance, a frequency comb (in GHz) 529.23, 1093.8, 1657.7, 2221.5, 2617.7, 2700.1, 2785.4 could be observed in the transmission dependence if the eigenstates are excited through an input prism at $\theta_1 = 67.58°$. The structure under consideration has a relatively big optical contrast and therefore at $\theta_1 > 40°$ the bands become extremely narrow and all the angle-frequency space of the upper panel is filled by narrow lines of bands and local states of D and S types. A complex game of transformations between the local states is visually shown in [21]. With the growth of $d_D$, the local defect modes exhibit a rather complex behaviour amid a relatively weak transformation of band and surfacestates.

A general description of reflection is given in the lower panel of Figure 2a for p-polarization where the reflecting properties of the (Si/PE)$_5$/D (Si/PE)$_5$/Si photonic resonator are depicted in 16-grade color scale. The existing inside the first reflection window resonance **#1** occupies the frequency interval (590.76, 692.75) GHz. This resonance is generated by three close local states S-S-D above the boundary of the TIR region and open angular area. Therefore, it has a triple structure at angles of whispering incidence $\theta_l > 89.4°$. This triple line transforms into a unitary resonant curve of transmission at $\theta_l < 89.4°$. The first in time defect caused resonances **#3** detaches from the bottom of fourth transmission window when the width of defect layer reaches $\approx 2.5$ μm. The next one **#1** detaches from the bottom of second transmission window when the width of defect layer reaches $\approx 8$ μm [21]. If the number of periods growths the triple zone is shifted closer to 90°. At N=30, the resonant curve virtually "feels" the local eigenstates of the TIR region beginning with 89.992° taking the corresponding triple form. The following resonant lines **#2**, **#3**, **#4** have the same triple-unitary structure and are situated in the next three reflection windows if we will consider a wider frequency interval (0, 3) THz. The shape of a resonant transmission curve (HWHM) and its frequency interval are important in our study. If the window of reflection is wider than the frequency interval of the situated there resonant curve then the latter, like a dispersive prism in optics, can be used to transform the frequency distribution of the beam density of intensity into the corresponding angular distribution. The narrower is the curve of transmission resonance - the sharper should be the resulting angular distribution. Our calculations show that for any resonator based on Si-PE and Si-PP pairs independently of geometry only first reflection window can cover the frequency interval of the resonance **#1**. The reason is that the intrinsic contrastivity of the structure is not high enough. Nevertheless, in the case of p-polarization, the more contrastive Si-air based structures have even a worse characteristic of overlapping the frequency intervals because of the Brewster effect. It is worth to note the surface states are principally absent if we consider a resonator based on a pneumatic photonic crystal PE→Air and therefore the trident structure of transmission peaks at whispering incidence vanishes even in the case of p-polarization. The parametric evolution of resonances on the map of reflection is presented in [21]. With the growth of the width of defect layer, the corresponding resonances consequently detaching from the bottom of upper window of transmission and, after a travel through the space of reflection window, join the top of lower transmission window. Therefore, only **#1a** and **#1b** resonances are present in the first window of reflection at $d_D=200$ μm [21].

In Figure 2b, top panel, the bandgap structure of s-polarized field inside the resonator in angular interval $\theta_1 \in (17.58°, 57.58°)$ is presented. In general, the structure looks more contrastive in s-polarized light than for p-polarization: gaps are relatively wider. The s-polarized band modes exhibit the same paired structure with transformation at $\theta_1 > 28.30°$ when sinusoidal standing waves inside PE layers become hyperbolical (highlighted with color). The point of defect local state detachment from the bottom of the first band lies much higher for s-polarization ($\theta_1 \approx 44.49°$, left vertical arrow). A position of surface state S detachment at $\theta_1 \approx 51°$ is marked by the right vertical arrow. At small angles $17.58° < \theta_1 < 51°$, only defect local states arise: the second D-branch begins at 17.58° in the middle of the first gap at given resonator parameters. Our evaluation shows that this local state starts its detachment from the bottom of the second band (inclined arrow) when the ratio $d_D/d_{Si}$ reaches 0.055 to occupy the shown position at $d_D=d_{Si}$ (Figure 2b, upper panel). In the case of a thick defect layer $d_D > 3.6 d_{Si}$, four local D-states detach from the bottom of each band. As long as the bands are essentially converged in the upper part of the Figure 2b, top panel, the total angle-frequency space at $\theta_1 > 60°$ becomes filled by a set of inclined narrow lines of bands and local states. At $\theta_1 > 26.29°$, the transmitted sinusoidal s-modes similarly to p-modes became exponentially decreasing in PE layers and transmitted standing waves transform into the waveguide type.

In Figure 2b, lower panel, the angle-frequency color map of reflection is plotted in a 16-grade color scale for s-polarized. The existing inside the first reflection window resonance **#1** occupies the frequency interval (590.76, 643.83) GHz. It is generated by three close local states S-S-D at the boundary of TIR and open angular area and therefore has a triple structure at angles of whispering incidence $\theta_1 > 89.4°$. The trident form of the resonant transmission peak transforms into a Gaussian-like form at $\theta_1 < 89.4°$. The following resonant curves **#2, #3, #4** are situated in the next reflection windows of the interval (0, 3) THz though the areas of windows do not cover this curves totally. Therefore they are not so interesting in the framework of our study. One more peculiarity of s-polarization is that surface local states are absent for small angles of the TIR region $\theta_1 \approx 17.58°$ and only defect states are present (Figure 2b, upper panel). Therefore their continuation into the area of external incidence where transmitted waves pass through the resonator has a view of a narrow resonant Gaussian-type peak in the angle-frequency space against the background of perfect reflection in the total interval of incident angles $0° < \theta_1 < 90°$. A useful role of this circumstance will be considered below.

It was been mentioned above, the presence of defect would conditionally cut the PhCr into two parts possessing the same spectrum. Therefore, the resonator eigenstates form pairs which become closer with increasing the thickness of the defect layer. There exists in quantum mechanics a prohibition of eigenstates degeneration in a 1D system [22]. At the same time, this effect can be observed, for example, in spectrum of two interacting photonic crystal islands [13]. Therefore the considered above eigenproblem of a defect containing photonic resonator can be treated as a macroscopic manifestation of the quantum mechanical nondegeneracy effect occurring due to the wavelike nature of subatomic particles.

The spectrum of the $(Si/PP)_n/D(Si/PP)_n/Si$ photonic resonators has also been calculated. Being of the same form, it is shifted to blue side beginning with **0** at the left part of Figure 2 to approximately 20 GHz at the right side of frequency interval. The results obtained for $(PP/Air)_n/D$ $(PP/Air)_n/PP$ and $(PE/Air)_n/D$ $(PE/Air)_n/PE$ resonators show that the needed overlapping of resonant curve **#1** by the window of perfect reflection is much worse than that demonstrated in the case of $(Si/PP)_n/D$ $(Si/PP)_n/Si$, $(Si/PE)_n/D$ $(Si/PE)_n/Si$ and $(Si/Air)_n/D$ $(Si/Air)_n/Si$ structures.

The chosen resonator geometry is matched to the THz frequency region. Nevertheless, the scaling rule is acting for such kind hierarchical structures and the results obtained remain absolutely the same if a parametric transformation is made: $\upsilon \rightarrow q\upsilon$, $d_1 \rightarrow d_1/q$, $d_2 \rightarrow d_2/q$, $d_D \rightarrow d_D/q$, where q is the scaling coefficient. It means that spectrum and map of reflection calculated, by instance, for geometry $d_1=4$ μm, $d_2=7$ μm, $d_D=4$ μm (q=10) coincide with that presented in Figure 2a, 2b, they only expanded inside the frequency interval (0, 10) THz. This circumstance opens an unique possibility to design and fabricate the resonators with predefined properties using some technologically simple testing structures.

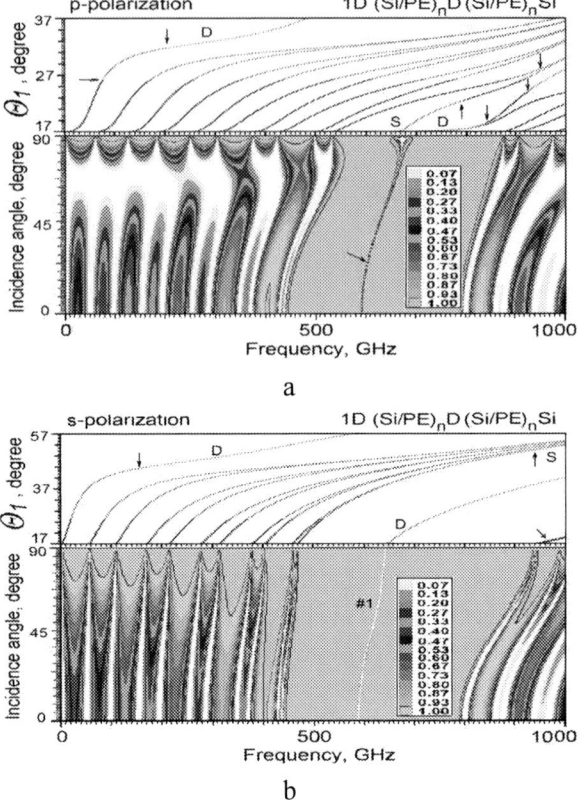

**Figure 2.** Spectrum of the 10 period $(Si/PE)_5/D (Si/PE)_5/Si$ photonic resonator with period 110 μm, PE voids size 70 μm. $d_{Si}=40$ μm a central silicon defect D of thickness 40 μm. (a) p-polarization. Top panel, angle interval $17.58° - 37.58°$: Bandgap structure and local modes Si/PE photonic crystal containing a central defect layer. Angle of incidence $\theta_1$ inside the Si material. Vertical arrows, points of local mode detaching/confluence. Horizontal arrow, angle of transmitted eigenmodes transformation into the waveguide type. D, defect local mode, S, surface mode. Lower panel: Angle-frequency diagram. External incident angle $\theta_l$. Notation #1 enumerates the defect caused transmission resonance. Inclined arrow, section of curve #1 at angle 22.87. Inset: scale of colors. (b) s-polarization. Top panel, angle interval $17.58° - 57.58°$: Bandgap structure and local modes of the 10 period Si/PE defected photonic of thickness 40 μm; $d_{Si}=40$ μm, period 110 μm, PE voids size 70 μm. Angle of incidence $\theta_1$ inside the Si material. Inside the interval $(0 - 1000)$ GHz two bands with first of them containing 10 resonator eigenmodes and from 1 to 3 local states detached from bands. Vertical arrows, points of detaching. Horizontal arrow, angle of reconstruction of transmitted eigenmodes to the waveguide type. Inclined arrows, defect local states. Bottom panel: Angle-frequency diagram. External incident angle $\theta_l$. Notation #1enumerates the defect caused transmission resonances.

## Shape and HWHM of the Defect Resonances

An extremely thin transmission lines against the background of perfect reflection have considered above the Si-PE and Si-PP resonators based on the 110 μm period PhCr. A total covering of the resonance curve by reflection windows may serve as an instrument for filtering and analysing the structure of radiation. This covering is observed only for resonance **#1**. It means that if the radiation of a monochromatic source with frequency $\upsilon_s$ matches the above mentioned interval (590.76, 692.75) GHz then it should be perfectly reflected for all angles but the critical one $\theta_{cr}=\varphi_1(\upsilon_s)$, where $\varphi_1(\upsilon)$ is the curve of resonance **#1** equation in the angle-frequency plane. In Figure 3, the transmission resonance **#1** is considered and its shape vs number of periods N dependence is presented for a $(Si/PE)_n/D\ (Si/PE)_n/Si$ photonic resonator at various **n** at normal incidence. Both p- and s-polarized resonances exhibit a blue shift of the frequency with a growth of N. The position $\Delta\upsilon = 0$ corresponds to $\upsilon_0 = 590.8030$ GHz in the case of p-polarization and $\upsilon_0 = 603.0675$ GHz for the s-polarized wave. In Figure 3a, the p-polarized transmission resonance **#1** at normal incidence is considered: curves 1, 2, 3, 4 for N=12, 14, 16, 18, correspondingly. The peak position relative to $\upsilon_0$ slightly depends on the number of periods (in MHz): +3.0168, +4.5143, +4.8247, +4.8890 in contrast to the HWHM which undergoes a strong influence of the number of periods (in KHz): 3842.15, 967.45, 200.55, 41.55 for N=12, 14, 16, 18, correspondingly.

The peak last half-width magnitude at N=18 is of $10^{-8}$ from the frequency that allows metrological aspects of THz radiation using relatively simple devices. The growth of the number of periods leads to further narrowing of the peak: N=20, HWHM=8.600 KHz, N=22, HWHM=1.785 KHz, N=24, HWHM=0.371 KHz. Thus, a 24-period device of approximate sizes 2.5mm×10mm×10mm allows one to operate with the $10^{th}$ digit of frequency magnitude of the THz source. In Figure 3b, the s-polarized transmission resonance **#1** at normal incidence is considered: curves 1, 2, 3, 4 for N=14, 16, 18, 20, correspondingly. The peak position relatively $\upsilon_0 = 603.0675$ GHz is slightly shifted depending on the number of periods (in MHz): +1.207, +1.437, +1.482, +1.490 for N=14, 16, 18, 20, correspondingly. The peak half-width strongly depends on the number of periods (in KHz): 540.53, 104.42, 20.175 and 3.9000 for N=14, 16, 18, 20, correspondingly. The peak half-width is as small as 28.125 Hz at N=26, it is of $5 \cdot 10^{-11}$ from the peak frequency 603.068992381 GHz.

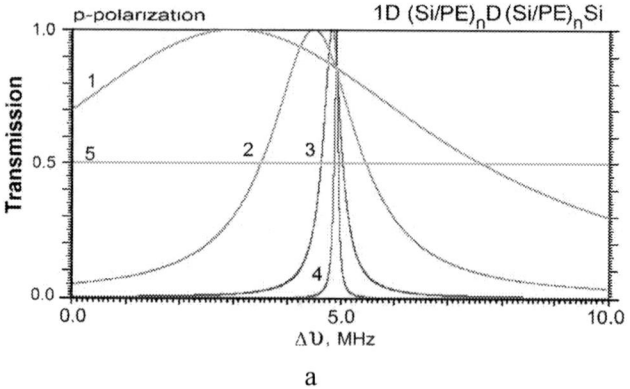

a

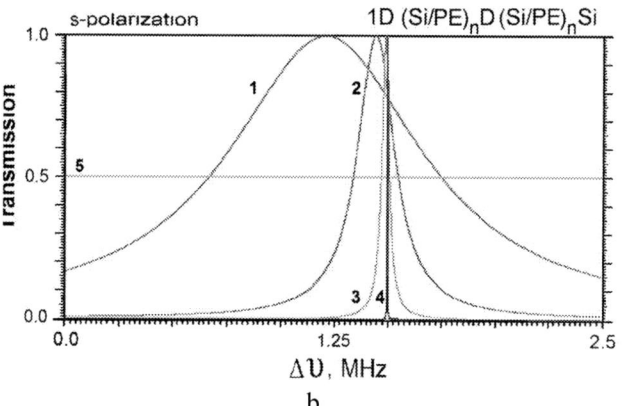

b

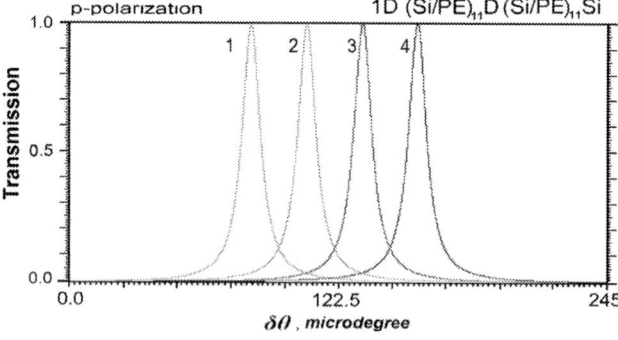

c

Figure 3. (Continued)

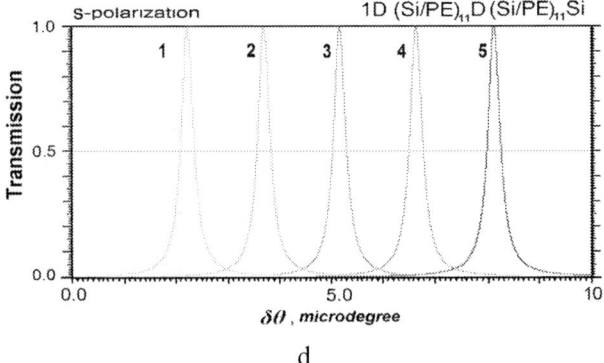

d

**Figure 3.** Calculated shape of transmission peaks. (Si/PE)$_n$/D (Si/PE)$_n$/Si photonic resonator. n=N/2. The resonance #1 shape at normal incidence. (a) p-polarization, resonant curve shape by frequency. Curves 1-4, at N = 12, 14, 16. 18, correspondingly; (b) s-polarization, frequency shape at θ=30°. Curves 1-4, at N = 14, 16. 18, 20, correspondingly. Curve 5, (R+T)/2 line in both cases. (c) p-polarization, angular shape. N =22. Curves 1 - 4, section of #1 (Figure 2a, lower panel) at υ = 599.99998, 600. 600.00002, 600.00004 (GHz), correspondingly. (d) s-polarization, angular shape. N =22. Curves 1 - 5, section of #1 (Figure 2b, lower panel) at υ = 599.99998, 599.99999, 600. 600.00001, 600.00002 (GHz), correspondingly.

The shape of the resonance curve in angular space has also meaning for our study. A section along the angular axis of the curve **#1** for p-polarized radiation in a 22 periodic central defect containing resonator is presented in Figure 3c for several suitable frequencies 20 KHz different from each other (in GHz): 599.99998, 600, 600.00002, 600.00004. Though the angular shifts are measured in microdegree units the peaks are not overlapping and angular distance between them is about 24.5 μkDeg, i.e., the derivative dθ/dυ≈1.23·10$^{-7}$ Deg/KHz. This means that the proposed structure forms narrow peaks that essentially divide the directions of different beam frequencies differing 20 KHz from each other. The graduating value of angle is taken φ$_0$(υ) = 22.865955° (Figure 3a, lower panel, inclined arrow). The angular shape of the s-polarized resonant curve is shown in Figure 3d for five frequencies (in GHz): 599.99998, 599.99999, 600, 600.00001 and 600.00002. The corresponding peaks are not overlapping, the distance between them is about 24.5 μkDeg, i.e., the derivative describing inclination equals to 1.5·10$^{-7}$ Deg/KHz. The appropriate graduating angle is φ$_0$(υ) = 25.73685°.

## Collimation Effect in a Planar Resonator Structure with a Central Defect Layer

One of the promising applications of symmetrically defected resonators is an opportunity to collimate radiation of a source with bad directional characteristic. An example was discussed in [23] to study how the existing sharp peaks of transmission in a metalized PhCr resonator can be used to control the divergence of the incident beam. Here we develop the approach presented there considering the lossless PhCr structures to improve the collimating properties of device.

The system under consideration is shown in Figure 4a. A monochromatic diverged incident beam $J_l$ passes from the source S into the resonator $(Si/PE)_n/D(Si/PE)_n/Si$ through an electromagnetic field reservoir (EFR). The latter allows one in a lossless process to concentrate inside the radiation of different directions with the angular white noise distribution. The source frequency should be matched with the resonant curve **#1** interval (590.76, 692.75) GHz if the above mentioned geometrical sizes of resonator are chosen. Depending on the quality of the source, the initial beam divergence $\Delta\theta$ may be large enough. The task is to collimate a diverging monochromatic beam. After interaction with resonator, any wave with a wavevector out of the transmission peak zone reflects back into the EFR space whereas waves inside the peak angular interval pass through the collimator with an significantly lowered divergence $\delta\theta$. We suppose that a stochastic angular distribution of field is created inside the EFR and distribution of power along the axis OX is homogeneous (Figure 4a). Owing to the perfect mirror walls of the reservoir the density of field W inside the EFR increases up to state of saturation when the energy flux of the output beam $J_r$ plus flux of losses become equal to energy flux of the input beam $J_l$. In [23], a 90 GHz beam being initially diverging in a wide angle interval has decreased its divergence up to 0.4° in the vicinity of the direction 22.5° after passing the metalized silicon-propylene resonator. An advantage of a metalized resonator is its ability to form narrow transmission resonance lines against the background of perfect reflection [12, 14]. The reason is the uniform mirror metallic reflection in the entire frequency interval. The disadvantage is the inevitable loss of power, which leads to a reduction in the output power of the THz searchlight. The considered above semiconductor-dielectric structures are characterized by the practically absent heat losses or scattering in the transparency region of materials. Therefore only the geometrical losses caused by escaping of radiation through the resonator faces are retained and as well as that inside the source of THz radiation.

We will take into account two possible channels of energy leakage: through the lateral sides of the PhCr and one more arising due to the reverse flow immediately through the source window A simplified kinetic equation for $W$ can be written as:

$$\frac{dW}{dt} = \frac{J_l S_l}{V} - \frac{cW}{V}(S_r \Omega_r f_r + S_l \Omega_l f_l), \qquad (4)$$

where $V$ is the volume of reservoir, $\Omega$ is spatial angle and $f_l, f_r$ are the integral characteristic of transmission through the corresponding window of the reservoir – left ($l$) or right ($r$). We suppose here that the intensities of radiation $J_l$ and $J_r$ reflected or transmitted through the metalized PhCr are proportional to the average density of energy $W$:

$$J_r = Wc \frac{\Omega_r f_r}{4\pi},$$

$$J_l' = Wc \frac{\Omega_l f_l}{4\pi} \qquad (5)$$

The characteristic time $\tau$ of the process of energy accumulation depends on the input and output sections $S_l$, $S_r$, the angle dependence of transmission for the chosen frequencies in the vicinity of *600 GHz* (shown by arrow in Figure 3a, lower panel), and volume $V$ of the EMW reservoir

$$\tau = \frac{V}{Wc(\Omega_r S_r f_r + \Omega_l S_l f_l)} \qquad (6)$$

Figure 4b shows the time-dependent intensity of the collimated output flow $J_r$ for the case the thickness of defect layer equals to $d_1$. The integral parameter of transmission $f_r$ represents geometrically area under the curve of transmission *1-R* shown in the Insertion. On account of perfect reflection and extremely narrow transmission curve width the calculated $f_r$ were of the order of $10^{-7} - 10^{-8}$ for s-polarization. The characteristic time $\tau$ of the process of energy accumulation also depends on the input and output sections $S_l$, $S_r$, output and input spatial angles, and volume $V$ of the EMW reservoir. In Figure

4b, time dependence of the collimated output flow $J_r$ is shown for the adopted parameters: $I_l$ =0.1 W/cm², $S_l$ =12.5 cm², $S_r$=12.5 cm², EMW reservoir volume $V$=1257cm³. The energy loss on the reservoir walls and the reverse flow through the source section $S_l$ is taken as 28%. As a result, we have a strongly collimated beam with the angular HWHM $\delta\theta$ close to 0.123 µDeg and output intensity $J_r$ =0.07142 W/cm². Our evaluations show that electromagnetic energy of saturation inside the electromagnetic box depends on the leakage through the windows. In the considered case, the accumulated inside the EMR energy approximately equals to 4.723 mJ and the saturation is reached during ≈35 ms with the characteristic time $\tau$≈3.778 ms. The resonant curve angular HWHM essentially depends on the structure optical contrastivity and the number of periods.

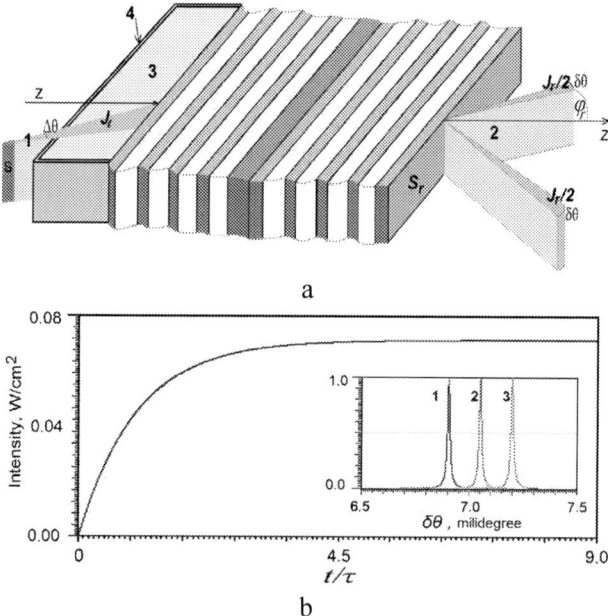

a

b

**Figure 4.** (a) (Si/PE)₄/D(Si/PE)₄/Si THz collimating system. 1, monochromatic diverged incident beam $J_l$; 2, collimated outgoing beams $J_r$/2, 3, EM field reservoir; 4, mirror walls of the reservoir; $\varphi_r$, output angle; $\Delta\theta$ and $\delta\theta$ are the input and output divergence of beam, correspondingly. (b) (Si/PE)₁₀/D(Si/PE)₁₀/Si resonator. Time dependence of the collimated flow $J_r$ for $\nu_l$=600GHz, $\tau$≈3.78 ms. Inset: Sharp peaks of angular dependencies of reflection at 600 GHz (curve 1), 600.0001 GHz (curve 2) and 600.0002 GHz (curve 3). Graduation angle is $\theta_0$≈25.730°. Reflection dip position of the curve 1 is $\theta_l$≈25.737° with divergence $\delta\theta$=0.123 µkDeg°.

We have considered in this section the case of a plane monochromatic beam having a wedge-like cylindrical radial divergence (Figure 4a). As the result of passing the resonator, two extremely collinear output beams arise at angles $+\varphi_r$ and $-\varphi_r$ in accordance with the curve of resonance **#1** (Figure 2b, lower panel) calculated for $N=20$. It is worth noting that both output beams are radiated by the all right surface of the resonator surface $S_r$. Both outgoing beams $J_r/2$ can be united using two plane mirrors at the some distance from the resonator. The collimation effect is expressed by the ratio between input and output beam divergences $\Delta\theta/\delta\theta$. Our evaluations show that this ratio may be bigger than $10^9$ ($N=22$, $\upsilon=600GHz$, $\varphi_r=15.01704881°$). If the angular distribution of incoming diverging beam has a conic symmetry the outgoing beams $+\varphi_r$ and $-\varphi_r$ transforms to a conic beam with homogeneous distribution of wavevectors at the polar angle $\varphi_r$. It is also possible to transform the outgoing beam into a collinear one using a corresponding remote conic mirror.

## Local Resonance Based Spectroscopy of Millimeter Waves

If a source radiates in the frequency interval occupied by the local resonance, the structure under consideration can serve as a prism which decomposes light depending on the wavelength by different refraction angles. We distinguish the p- and s-polarizations of radiation because the resonant curves of different polarizations coinciding for the normal incidence are going apart at the inclined incidence. Therefore, a mix of polarizations will give two (four) directions $\pm\varphi_{rs}(\upsilon)$ and $\pm\varphi_{rp}(\upsilon)$ of the outgoing monochromatic beam (Figure 4a). This effect can serve as a way to test the polarization of the source. One more possibility touches a novel way to investigate virtually the local modes of the TIR region in geometry of external incidence (Figure 1a) in the framework of the concept described in the previous Section. Though the TIR area is not attainable in geometry of external incidence an investigation of local states can be made with the use of a monochromatic source tunable in the interval of reflection window. In the case of s-polarization, a scanning of the resonant curve **#1** in the interval (590.76, 692.75) GHz will lead to arising a unitary extremely collimated beam passed through the system normally to the surface of the PhCr ($\varphi_{rs}(\upsilon)=0$). With the following growth of frequency, the passed collimated beam becomes double that corresponds to $\pm\varphi_{rs}(\upsilon)$ directions. In the vicinity of the upper end of resonant curve, the latter becomes triple that will lead to appearing of six collimated beams. Every one of them may be treated as a virtual image of surface or local eigenstates of the resonator obtained without the excitation of eigenstates inside the TIR region. In Figure 5a, shown is a calculated trident structure of the resonant curve at $\theta_1=89.999°$. The peaks in the sequence S-S-D are divided by the distance $\approx 90$ MHz. Thus, a conclusion can be made that local modes of the Si/PE)$_{15}$/D(Si/PE)$_{15}$/Si resonator begin at $\theta_1=17.58°$ inside the unattainable TIR region at 673.05 GHz, 673.14 GHz, and 673.22 GHz. The expressed triple structure of the p-polarized resonant curve at N=30 vanishes at $\theta_1<89.993°$. In the absence of accumulating electromagnetic box EFR, the same experiment will give only three directions of the outgoing beam with positive angles.

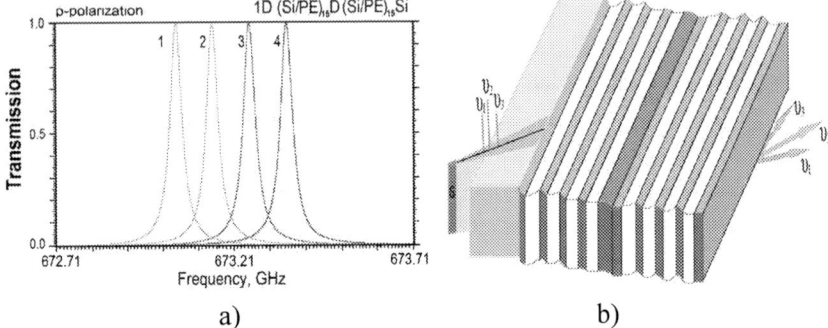

**Figure 5.** Spectroscopy of microwave radiation. (a) Si/PE)$_{15}$/D(Si/PE)$_{15}$/Si resonator; trident peak of transmission at whispering incidence; p-polarization; $\theta_1=89.999°$, S-S-D is the sequence of surface or defect type resonances. (b) A plane resonator can serve as a prism. S, source; $\upsilon_i$, i=1, 2, 3, beam frequencies; negative angles of beams are not shown; intensity is shown by bold vertical legs.

## Spectroscopy and Metrology of a THZ Source

Let the angle-frequency structure of the outgoing beam has a view of a narrow band with separated spectral peaks like it is shown in Figure 5b. The frequency dependence of intensity $J(\upsilon)$ contains a few peaks $\upsilon_i$:

$$J(v) = \int_v dv \sum_i \delta_i (v - v_i), \qquad (7)$$

where the functions $\delta_i$ describe the shape of the source generated peaks. The full intensity of the beam is given by integration on the entire interval of frequencies. If the peaks $\upsilon_i$ are matched with the operating frequency area of the collimator (590, 645) GHz for the system under consideration, the resonator begin to play a role of a spectrometer due to the presence of the EMR. The interaction of a beam with the resonator leads to a weak transmission at allowed angles of peaks $\upsilon_i$ crossing with the resonant curve of transmission. For any other incident angles the wave is ideally reflected. In the absence of the EMR, the extremely weak outgoing beam is split onto several corresponding to $\upsilon_i$ directions. But if the reflected radiation is remained inside the EMR, the accumulation of electromagnetic energy occurs inside the box up to a moment when the ingoing beam power becomes equal to the sum of outgoing power and power losses. The allowed angles of the collimated output beams are determined by the equation of the calibrated resonant line #1: $\theta_1(\upsilon)$. Therefore, such a kind of spectrometer points radiation of different frequencies to different directions, i.e., the frequency distribution of intensity is projected onto the angular one. The angular width of an outgoing beam along the direction $\theta_1$ contains two terms taking into account the finite width of transmission curve:

$$\delta\theta_i(v) = \left(\frac{\partial \theta_i}{\partial v}\right)_{v_i} \delta v + \delta\theta_i(v_i) ,\qquad(8)$$

and the angle-frequency dependence of the frequency density of intensity $j_r$ has a view:

$$j_r(v,\theta_r) = \theta_1(v) \sum_i \delta_i(v - v_i) \qquad(9)$$

The integration of $\mathbf{j_r}$ by frequency gives the output intensity in the direction of angle $\theta_r$. If the shape of peaks $\delta_i$ is close to the Dirac function, i.e., the spectrum is contained of the separated peaks $\mathbf{i}$, the spectrometer issues a system of collinear beams $\delta_i$, that correspond to frequencies $\upsilon_i$ with intensities proportional to heights of the density of intensity $\mathbf{j_r}$ with the angle of direction $\theta_i(\upsilon_i)$. As well as in the previous paragraph, we are considering here a resonator with infinite sizes in YX plane. If the resonator has finite lateral sizes, weak accompanying beams may arise with negative angles $-\theta_i(\upsilon_i)$. The reason of accompanying beams is reflection from the corresponding side ends of the structure.

A layered photonic structure with a central defect is able to form the extremely narrow transmission resonances. An example was considered above: the resonant curve #1 at $\upsilon$=606045.738 MHz (Figure 2b, lower panel) has the HWHM about 3.5 KHz at the incident angle 33.764° for a 20-period Si-PE resonator, i.e., a change occurs in the 9$^{th}$ digit of the frequency magnitude. If N=10, the change touches the 5$^{th}$ digit of the frequency magnitude. This means that velocity of narrowing is approximately 5 digits per 10 periods. The angular width at this frequency is also extremely small: the HWHM is approximately 0.0681 μkDeg. As a consequence, a minimal variation of the incident angle changes sufficiently the transmission in dependence on the frequency, resonant curve shape and its width.

**Resonant Transmission Based Filtering**

The sharp resonances of transmission looking like thin lines onto the angle-frequency diagram also give an opportunity to organize tuneable ultra narrow filtering of a collimated terahertz beam. The filter spectral width depends on the resonator quality, chosen frequency interval and the incident angle of a collimated beam. A resonator based on materials of higher contrast like boron contained silicon has issues more wide windows of reflection and sharper local states with correspondingly narrower HWHM. One can predict two cases of possible application. (1) If the frequency band of the collimated source matches one of the considered in above windows of reflection (Figure 2, lower panels) at the corresponding direction of incidence then only frequencies inside the peak of transmission will pass through the resonator whereas the rest will be reflected. It means a sufficient weakening of the output signal because the resonator bandpass is narrow. The use of the discussed above stochastic EMW reservoir allows to enhance the output beam intensity regardless the narrow angular area of the resonator. (2) Another possible application corresponds to the case when a source generates a collimated but instable monochromatic frequency. When the instable source scans an interval which includes the transmission resonance frequency then only the resonant frequency will be transmitted through the resonator; therefore the filter of that kind can play the role of a frequency stabilizer.

## Conclusion

The origin of transmission and absorption electromagnetic resonances is vividly discussed in literature. Although they are all bound by the resonant nature, they can nevertheless be distributed over several types. In the central defect containing lossless structures, the nature of resonance is bound with accumulation of the constructive interference at the proper frequencies and angles when a narrow band of perfect transmission arises against the background of perfect reflection. This is confirmed by an essential dependence of resonant peaks sharpness on the number of periods independently of the incident angle (Figure 3).

In the case of a metalized resonator, two types of resonance can be observed. One of them – Fano resonance is inherent to processes of energy loss on metals including light transmission through the metalized PhCr. As it is mentioned in [24] the induced currents in a conducting medium lead to extra losses of energy at the proper resonant frequencies due to the interaction of charge carriers with phonons. The reflection dips have an asymmetric shape for this case. This asymmetry was observed in [12], where the reflection spectra of metalized plane resonators were studied both for p- and s-polarizations. It was found that in the case of s-polarization, the resonances have a view of Fano type for any incident angles. As to the p-polarization, the resonance exhibits a mixed nature: at the high enough longitudinal momentum of the incident wave, a surface plasmon spike in reflection arises when condition of resonance is satisfied. If the incident angle is not exceeding the Brewster value the wells in reflection exhibit properties of the Fano resonance. Both these mechanisms describe some kind of resonant cancellation of reflection in a metal-resonator system. A similar resonant cancellation phenomenon of absorption which occurs in coupled optical resonators due to mode splitting and destructive interference was demonstrated in [22]. The photochemical mechanism of hole burning in reflection spectra is also long ago known [25].

Here, we have studied an effect of extraordinary narrow resonant transmission peaks on the background of perfect reflection arising in a planar central defect containing photonic resonator. A tight correlation between the resonator eigenstates existing inside the total intrinsic reflection region and unusual manifestations of THz transmission spectra in centimeter and millimeter wavelength range was considered. The phenomenon of resonance is expressed as the thin transmission curves in the reflection windows with extremely small angle and frequency widths that can be less than $10^{-9}$ of the corresponding angle and frequency magnitude depending on the number of periods. This allows one to form the collimated beam with the divergence measured in fraction of a microdegree. It means that a thin outgoing beam of radius 10 cm widens to 20 cm radius at a few tens kilometres away. This gives an opportunity to form long and stable channels of communication in the THz frequency range. It was also shown that a plane resonator containing a central defect transforms the frequency like a prism can divide peaks into the outgoing-transmitted beams of various directions. This opens the way for precision measurements of angle and frequency distribution of THz radiation. It was proposed here to use the arising sharp peaks of transmission for aims of spectroscopy and metrology. A way of spectroscopy was proposed that is based on the conception of accumulating reservoir of electromagnetic field.

# References

[1]  Zhang, X. C., Xu, J. (2010). Introduction to THz Wave Photonics. Springer Science+Business Media, LLC.
[2]  Lova, P., Manfredi, G., Comoretto, D. (2018. Advances in Functional Solution Processed Planar 1D Photonic Crystals. *Adv. Optical Mater.*, 6, 1800730.
[3]  Tonouchi, M. (2007. Cutting-edge THz technology. *Nature Photonics*, 1, 97–105.
[4]  Dhillon, S. S. et al. (2017). The 2017 terahertz science and technology roadmap. *J. Phys. D: Appl. Phys.*, 50: 043001.
[5]  Hesler, J., Prasankumar, R., Tignon, J. (2019). Advances in terahertz solid-state physics and devices. *J. Appl. Phys.*, 126, 110401.
[6]  Kyaw, C. et.al. (2020). Guided-mode resonances in flexible 2D terahertz photonic crystals. *Optica*, 7(5), 537-541.
[7]  Shi, X., Han, Z. (2017). Enhanced terahertz fingerprint detection with ultrahigh sensitivity using the cavity defect modes. *Scientific Reports*, 7, 13147.
[8]  Sizov, F. F. (2017). Infrared and terahertz in biomedicine. *Semiconductor Physics, Quantum Electronics & Optoelectronics*, 20, 273.
[9]  Wu, C. J., and Wang, Z. H. (2010). Properties of defect modes in one-dimensional photonic crystals. *Progress In Electromagnetics Research*, 103, 169–184.
[10] Jena, S., Tokas, R. B., Thakur, S., and Udupa, D. V. (2021). Thermally tunable terahertz omnidirectional photonic bandgap and defect mode in 1D photonic crystals containing moderately doped semiconductor. *Physica E: Low-dimensional Systems and Nanostructures*, 126, 114477.
[11] Chang, Y. H., Jhu, Y. Y., Wu, C. J. (2011). Temperature dependence of defect mode in a defective photonic crystal. *Opt.Commun.*, 285, 1501–1504.
[12] Glushko, E. Ya. (2020). Mixed Fano-SP resonant absorption of THz electromagnetic waves in a photonic resonator contacting with a metal film. *Physics Letters*, A 384(23), 126564.

[13] Glushko, E. Ya. (2017). Island-Kind 2D Photonic Crystal Resonator. *Ukr. Phys. J.*, 62(11), 945–952.
[14] Glushko, E. Ya. (2021). Induced resonant electromagnetic piercing in metalized photonic crystal structures. *Optik*, 231, 166502.
[15] Zaki, S. E., Mehaney, A., Hassanein, H. M., and Aly, A. H. (2020. Fano resonance based defected 1D phononic crystal for highly sensitive gas sensing applications. *Scientific Reports*, #10, 17979.
[16] Glushko, E. Ya. (2005). All-optical signal processing in photonic structures with nonlinearity. *Opt.Commun.*, 247, #4-6, 275–280.
[17] Glushko, E. Ya. (2008). Logical gates on trapped modes in photonic crystals with nonlinear coating. *Proc. SPIE* 6903, 69030G -69030G-10.
[18] Sanjuan, F., and Tocho, J. O. (2012). Optical properties of silicon, sapphire, silica and glass in the Terahertz range. *Latin America Optics and Photonics Conference*, paper LT4C.1.
[19] Treshin, I. V., Klimov, V. V., Melentiev, P. N., and Balykin, V. I. (2013). Optical Tamm state and extraordinary light transmission through a nanoaperture. *Phys. Rev. A*, 88, 023832.
[20] Kaliteevski, M. et.al. (2007). Tamm plasmon-polaritons: Possible electromagnetic states at the interface of a metal and a dielectric Bragg mirror. *Phys. Rev. B*, 76, 165415.
[21] Glushko, E. Ya. (2022). https://archive.org/details/modes-resonances-02.
[22] Landau, L. D., and Lifshitz, E. M. (1977). Quantum mechanics. Nonrelativistic theory, vol. 3. Pergamon Press, New York.
[23] Glushko, E. Ya. (2021). Collimation Effect on THz Transmission Resonances in Metalized and Defected Photonic Structures. in *Research Trends and Challenges in Physical Science,* edited by Prof. Shi-Hai Dong, vol. 1, Chapter 7, 83-89. B P International UK.
[24] Falkovsky, L. A. (1995). Resonance Fano in a System of Interacting Electrons and Phonons. *Pisma JETP*, 62, 227 - 230.
[25] Horie, K. (1991). Photochemical Hole Burning in Optical Materials, The Centennial Memorial Issue of The Ceramic Society of Japan, 99, 912-922.

## Chapter 3

# Antenna Array of Upgraded Quadruple Sector Emitters for Wireless 4G Networks

## Andrii Karpenko*

The Interdepartmental Scientific and Educational Physics and Technical Center
at MES and NAS of Ukraine, Odessa I.I. Mechnikov National University,
Odessa, Ukraine

### Abstract

Based on the "Quados" radiators, developed an optimized antenna array with a gain of 17÷12,5 dBi in the frequency band 3 (1710÷1880 GHz), designed for efficient operation with 4G LTE-1800 mobile operators in areas of weak signal strength. The following minimum values were selected as optimization criteria: 1. Standing wave coefficient in the operating frequency range; 2. Weight and sail of the antenna; 3. The number of matching devices.

**Keywords**: 4G antenna, wireless internet, cellular communication

### Introduction

The Internet has become firmly embedded in almost every sphere of human activity. New ways of providing services, professions and forms of labor relations are rapidly emerging. The need for Internet communications increases every year. In this regard, the task of providing a reliable high-speed Internet connection is becoming even more urgent.

---

* Andrii Karpenko, MS; Corresponding Author's Email: karpenko@onu.edu.ua.

In: Electromagnetic Waves
Editor: Manuel B. Hutchinson
ISBN: 979-8-88697-254-2
© 2022 Nova Science Publishers, Inc.

In settlements located at a great distance from base stations and in areas of uncertain 4G signal reception, there is a problem of dependence of communication quality on weather conditions, speed reduction or total loss of wireless Internet connection. The reason for this is the lack of direct visibility between the subscriber and the base station, the long distance between them, as well as the peculiarities of terrain (hollows, ravines, mountains, woodlands, etc.), affecting the conditions of radio waves propagation. To solve this problem, directional antennas are used, having a high gain, which should be installed at a high altitude and oriented in the direction of the nearest base station. Such antennas must have a low weight, high mechanical strength, resistance to temperature variations, precipitation and UV radiation, as well as low sailing to withstand winds.

One common solution is the stationary antenna of type "Yagi-Uda" [1, 2]. The calculated gain limit of such an antenna is 19 dBi. However, the efficiency of the antenna "Yagi-Uda" decreases at HHF frequencies. Foremost, there is a need for its careful tuning. And, the tuning is somewhat similar, but exceeds in complexity the tuning of the multiline band pass filter. Industrial antennas must also be tuned at the stands before sale. To increase the gain, it is necessary to add passive elements to the antenna - directors.

Typically, Yagi-Uda antennas contain a Pistolcors loop vibrator. The loop vibrator itself has an input impedance of about 300ohms and coordinates well with a 75ohm coaxial cable feeder by using a half-wave loop. The loop reduces the input impedance by a factor of 4, from 300 to 75ohms, and provides symmetry. When passive elements are added to the loop vibrator, the input resistance of the antenna is greatly reduced. Thus, the input resistance of the five-element antenna, depending on its size, may be in the range of 40...120Ohms. Being additionally reduced 4 times by the half-wave loop, it falls to 10... 30Ohm, which leads to a sharp mismatch of the antenna with the feeder. Due to the reflection of a significant part of the received signal energy and its radiation back into space, the antenna gain is significantly reduced.

In addition to the five-element ones, the dimensions of seven-element, eleven-element "Yagi-Uda" antennas, as well as with an even greater number of elements have been developed and published in some literature sources [3]. Without careful adjustment, such antennas, even made exactly according to the drawings, have poor characteristics. In addition, as the number of elements increases, the antenna bandwidth narrows. For example, the bandwidth of a seven element Yagi-Uda type antenna is approximately 5% of the frequency to which it is tuned.

An effective variant of the antenna "Yagi-Uda" is another common solution in the form of a disk antenna, in which all elements are made of metal disks [4, 5]. The gain of this type of antenna can reach 20 dBi. The radiating element here is not a loop vibrator, but a disk-shaped patch element. The advantage is the ability to connect two coaxial cables in certain points of the patch, providing simultaneous operation in two orthogonal polarizations, which is necessary for MIMO format data exchange. The choice of connection points on the patch affects the wave impedance of the antenna, which is in the range of 50-75 ohms, which makes it possible to do without wave impedance transformation and symmetry. However, setting up this type of antenna is no less complicated than in the classic "Yagi-Uda" antenna.

Another common solution is the use of antenna arrays based on patch elements [6-8]. Their advantage over the "Uda-Yaga" antennas is ease of tuning, lack of need for matching and symmetry with the coaxial reduction cable, as well as short-circuiting the input for direct current, which protects the modem from possible electrostatic breakdown. However, their large mass and salability are far superior to other types of antennas.

The author of this work has set himself the task of selecting and optimizing for 4G communication such a type of antenna, which satisfies the following requirements: 1. Minimum standing wave ratio (SWR) in the operating frequency range; 2. The minimum mass and sailing of the antenna field; 3. The minimum number of symmetrical and matching devices; 4. Maximum gain (Gain); 5. Mobility of design (possibility of transportation, fast assembly, disassembly, installation).

## Selection of the Prototype

The Quados sector antenna for 2.4GHz Wi-Fi networks [9], which was scaled to the 4G Band 3 band and modeled in the "MMANA-GAL" program, was chosen as the basis for solving the problem [10]. The wave impedance of the Quados is 200ohms at 2.4GHz. The author [10] solves the problem of matching with the coaxial cable with a wave impedance of 50ohms by using the classic half-wave loop, simultaneously performing the function of symmetry. However, when scaling the antenna for the desired range of 1710 ÷ 1880GHz in the computer model, there is a significant scatter of the wave impedance of the antenna - from 180 to 70ohms. As a result, at the output of the half-wave loop, the wave impedance is from 45 to 17.5Ohms. The resulting mismatch with the reduction cable increases the SWR to 3.8 at the end of the operating frequency range (Figure 1), which is unacceptable for the task at hand.

However, the Gain in the middle of the frequency range reaches a value of 17.5 dBi, which together with a low sail design was a strong argument for optimizing the antenna "Quados" to work in the range of 4G LTE wireless networks.

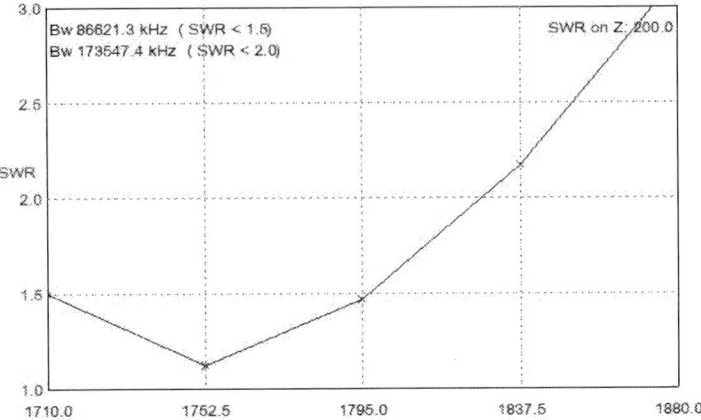

**Figure 1.** SWR of the Quados reflector-loaded half-wave loop antenna in the Band 3 frequency range.

## Solution Method

In order to achieve the goal, according to the established requirements, the "Quados" antenna was modernized according to the following algorithm:

a. The reflector is removed from the computer model of the antenna "Quados" (Figure 2).
b. By changing within small limits the geometric parameters of the antenna, the value of the antenna wave impedance in the middle of the working frequency range, close to 100Ohm, is achieved.
c. Two identical elements (stacks) of the antenna array are created from the obtained model, with their parallel feeding by segments of the symmetric long line with the wave impedance equal to 100Ohm (Figure 3). In this way, the matching with a 50 ohm reduction coaxial

cable is achieved, which eliminates the use of an impedance transformer.

A symmetrical long line with a wave impedance of 100Ohm is made from two sections of 50Ohm coaxial cable with electrically connected braids, or a two-core shielded cable with a wave impedance of 100Ohm is used (Figure 4). The cable braid is not connected anywhere.

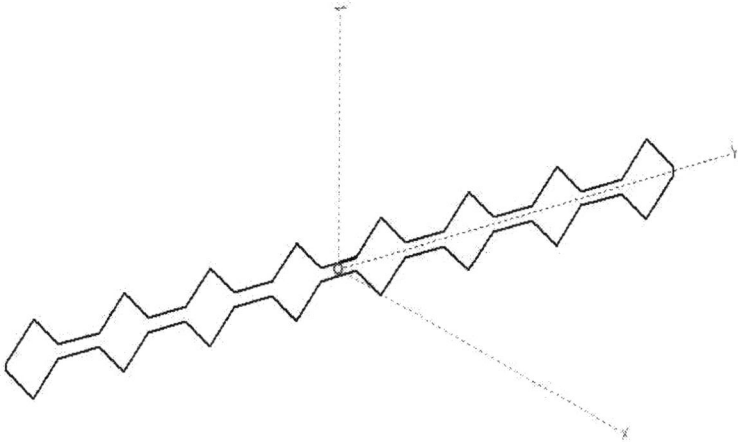

**Figure 2.** Quados antenna model without reflector in the Mmana-Gal program.

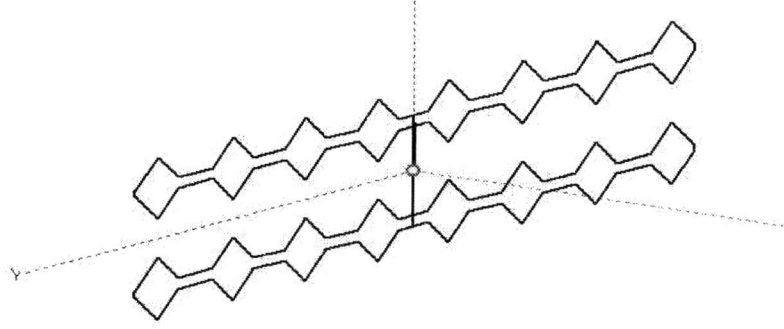

**Figure 3.** Model of the antenna array of 2 stacks of optimized "Quados" without a reflector in the program Mmana-Gal.

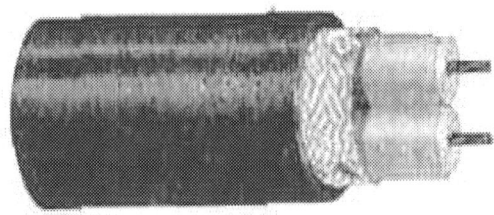

**Figure 4.** Symmetrical long line with shielding.

## Simulation Results. Discussion

In the process of scaling and optimizing the "Quados" antenna (Figure 2) for the 4G Band 3 frequency range, the gain at the center frequency of 1795 MHz reached a Gain = 12.65 dBi and the wave impedance was $\dot{Z} = 99.86 - j5$ Ohms. The smooth variation of the wave impedance in the Band 3 frequency range with a small drop in values satisfies the technical requirements of the task (Figure 5).

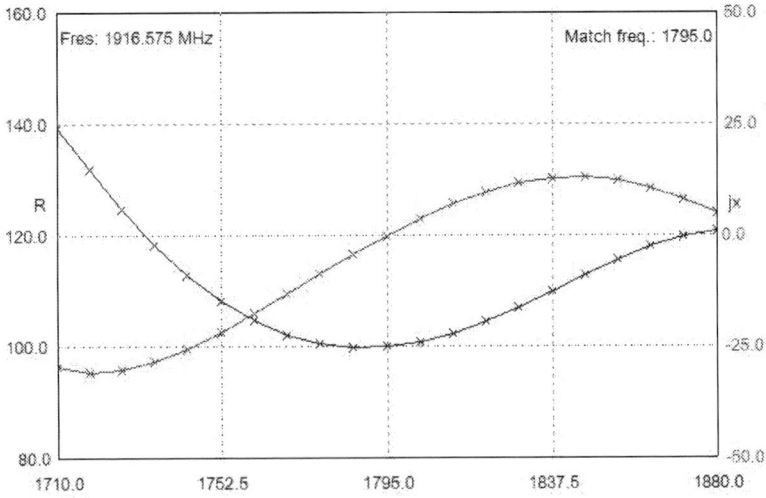

**Figure 5.** Active and reactive components of the wave impedance of a single optimized "Quados" antenna without a reflector in the Band 3 frequency range.

As a result of modeling and optimization of the antenna array (Figure 3) the following characteristics were obtained:

a. In the middle of the frequency range of 1795 MHz, the wave resistance at the feed point of the antenna array has an active character and is $\dot{Z} = 55 - j0.5$ Ohms. The smooth variation of $\dot{Z}$ with a small drop in values satisfies the technical requirements of the task and is within acceptable limits for modems designed for wireless networks (Figure 6).

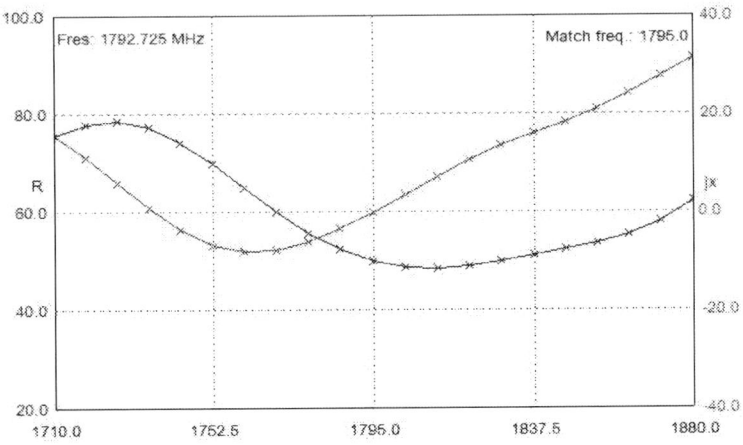

**Figure 6.** Active and reactive components of the wave impedance of the antenna array without a reflector in the Band 3 frequency range.

b. A symmetry device is necessary only at the point of connection to the reduction coaxial cable, because the elements of the antenna array, themselves being symmetrical antennas, are powered by two short segments of a symmetrical line. If a modem with a symmetrical antenna input is placed at the antenna array supply point, a symmetrical antenna input can be dispensed with.

c. The standing wave coefficient in the middle of the Band 3 frequency range is SWR=1.1, and reaches a maximum of SWR=1.8 at the upper limit of the frequency range (Figure 7).

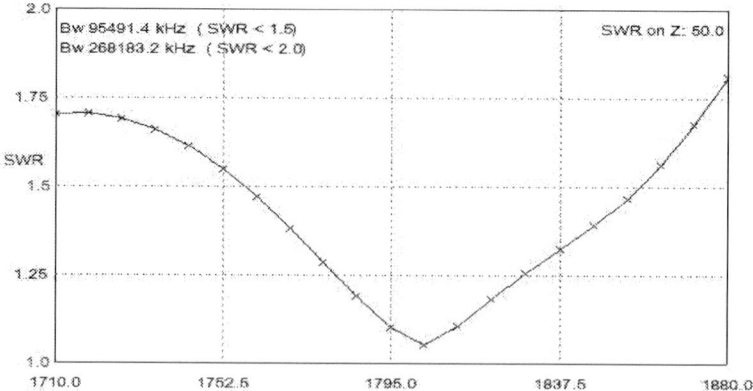

**Figure 7.** SWR of the Band 3 antenna array with a 50-ohm reduction cable.

Slightly better matching occurs with a 75ohm coaxial cable load. However, most modems have antenna inputs with a wave impedance of 50ohms and by connecting to them an antenna array with a 75-ohm reduction cable, the total SWR will increase and may exceed the values shown in (Figure 8).

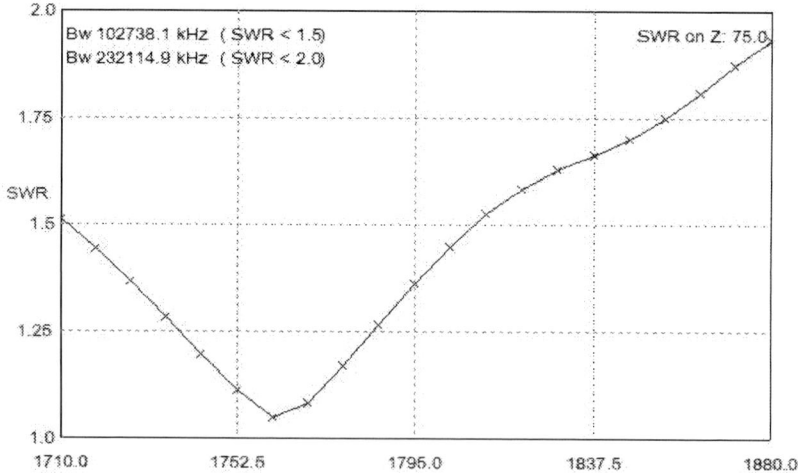

**Figure 8.** SWR of the Band 3 antenna array with a 75-ohm reduction cable.

Antenna Array of Upgraded Quadruple Sector Emitters ...       63

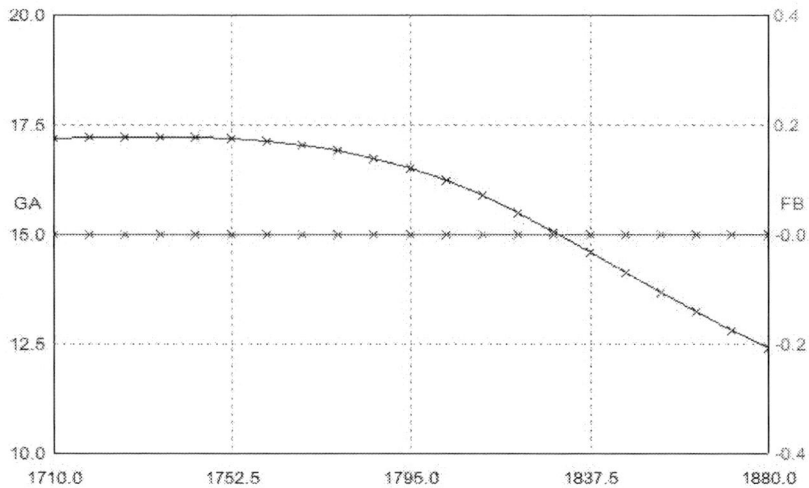

**Figure 9.** Gain of the antenna array at Band 3 frequencies.

d. The Gain at the beginning of Band 3 frequency range is 17 dBi, then decreases smoothly, taking values of 16.5 dBi in the middle and 12.5 dBi at the end of the frequency range (Figure 9). Despite this, the efficiency of the antenna array does not decrease due to the small SWR value in the frequency range, the minimization of which was the main focus. Since the antenna array in this model has no reflector, the forward/backward radiation power ratio is unity, i.e., 0 dBi (FB parameter) (Figure 9).

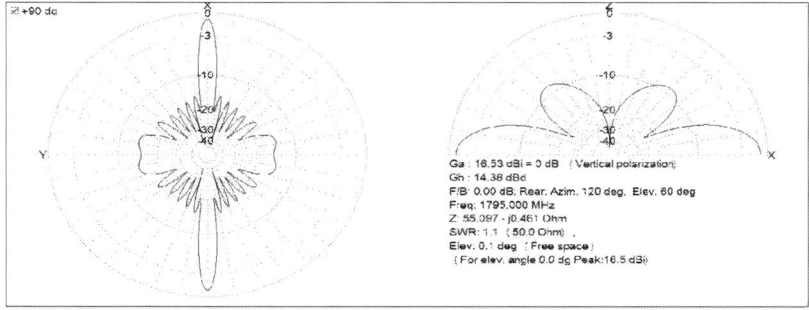

**Figure 10.** Antenna array directive diagram in the horizontal and vertical planes in the middle of the band 3 frequency range.

e. The directional diagram in the horizontal plane in the middle of the frequency band 3 has two main lobes, directed perpendicular to the plane of the antenna blade, mirroring in both directions (Figure 10). Throughout the entire angular range, a family of side lobes is observed, which is a fundamental disadvantage of antenna arrays.

## Conclusion

1. The model of the antenna array optimized for the range of 4G LTE wireless Internet, consisting of two stacks of antenna "Quados" with their parallel feeding sections of symmetrical long line with a wave impedance equal to 100 ohms meets the requirements, having the Gain values from 17 to 12.5 dBi in the frequency range (1710 ÷ 1880 GHz) and SWR values from 1.1 to 1.8.

2. Modeling of the antenna array with a reflector showed an increase of Gain by another 2 dBi and a significant increase in the FB parameter, but the SWR maximum reaches a value of 4.5, due to which the bandwidth of operating frequencies is four times narrower: from 200 MHz to 50 MHz, which eliminates the feasibility of using a reflector in this task.

## References

[1] Tenzin J, Rinzin C, Bdr. Samal P. (2019). 4-Element Yagi-Uda Antenna Offering Highest Gain at 2100 MHz. – *ETIC conference*.
[2] Mahardika RA, Broto S. (2019). Yagi Biquad Antenna Design for 4G LTE in 2100–2400 MHz Frequency Band.
[3] Ji-Hwan Hwang, Hyeong-pil Seo, Kyu-Lyong Cho. (2022). Trouble Shooting due to Short Link Loss of UAT using Yagi-Uda Antenna. *Journal of the Korea Academia-Industrial Cooperation Society*, Vol. 23, No. 4 pp. 68-75.
[4] Nachev I, Iliev IA. (2020). Simplified Design Methodology for Hybrid Antenna for S-band Application. *Microwave Review*. – Т. 26. – №. 2. – С. 14-18. http://www.mtt-serbia.org.rs/files/MWR/MwR-Vol26-No2/Vol26No2-2174-3-INachev.pdf.
[5] Sibruk LV, Fomenko NS. (2020). Вибір апаратури для забезпечення якісного мобільного зв'язку в сільській місцевості [Selection of Equipment to Ensure High-Quality Mobile Communication in Rural Areas]. *Проблеми Інформатизації та Управління [Problems of Informatization and Management]*. – Т. 2. – №. 64. – С. 54-58. https://doi.org/10.18372/2073-4751.64.15150.

[6] Wang H, Huang XB, Fang DG. (2008). A single layer wideband U-slot microstrip patch antenna array. *IEEE Antennas and Wireless Propagation Letters.* – Т. 7. – С. 9-12.
[7] Midasala V, Siddaiah P. (2016). Microstrip patch antenna array design to improve better gains. *Procedia Computer Science.* – Т. 85. – С. 401-409.
[8] Coulombe M, Koodiani SF, Caloz C. (2010). Compact elongated mushroom (EM)-EBG structure for enhancement of patch antenna array performances. *IEEE Transactions on Antennas and Propagation.* – Т. 58. – №. 4. – С. 1076-1086.
[9] Dobričić D, YU1AW. (2008). Quados Sector Antenna for 2.4 GHz WiFi. *Antenne X Online Issue,* No. 131.
[10] Antenna calculation and analysis software MMANA-GAL. http://gal-ana.de/basicmm/en/.

Chapter 4

# On the Scattering of Electromagnetic Waves by Bodies with Non-Coordinate Boundaries

### I. E. Pleshchinskaya and N. B. Pleshchinskii[*]
Kazan National Research Technological University,
Kazan Federal University, Russia

#### Abstract

This chapter explores the possibility of numerical solving scattering problems of electromagnetic wave by bodies with non-coordinate boundaries by reducing to infinite sets of linear algebraic equations relative to the coefficients of expansion of the unknown field in terms of eigen waves of coordinate domains in different coordinate systems. In Cartesian coordinates scattering problems of an electromagnetic wave by a non-coordinate media interface and by a conductive thin screen in a plane waveguide are considered. Two-dimensional and three-dimensional problems of diffraction of a plane wave by a periodically perturbed media interface in the open space are considered also. Expansions of the unknown field in terms of cylindrical and spherical waves are used in solving the problem

---

[*]Corresponding Author's Email: prosper7@yandex.ru

In: Electromagnetic Waves
Editor: Manuel B. Hutchinson
ISBN: 979-8-88697-254-2
© 2022 Nova Science Publishers, Inc.

of diffraction of an electromagnetic wave by a cylindrical rod with a non-coordinate boundary and a two-dimensional wave diffraction problem by the axis-symmetrical body. Some results of computational experiments are given.

**Keywords**: electromagnetic wave, scattering, diffraction, non-coordinate body

**AMS Subject Classification:** 78A45, 78M25.

# 1. Introduction

Scattering or diffraction problems of electromagnetic waves by dielectric and conductive bodies belong to the classical problems of electrodynamics [1]. In these problems we need to look for solutions of Maxwell equations outside the body and inside the dielectric body satisfying conjugation conditions on its boundary. The boundary conditions should be fulfilled on the boundary of the conductive body.

Analytical solutions of scattering problems can be obtained only in exceptional cases, as a rule, when the boundary of the body is a coordinate surface. Solutions of classical diffraction problems of a plane electromagnetic wave by a ball and by a round bar are well known (see, e.g., [2], ch. V).

These solutions are constructed in the form of expansions into a series in terms of spherical or cylindrical waves. Analytical expressions of the coefficients of expansions are easily derived from the conditions of conjugation of fields. The diffraction problem of an electromagnetic wave by a dielectric insert in a waveguide with metal walls is simply solved also, if the boundaries of the body are cross-sections of a waveguide.

A similar situation takes place when a plane electromagnetic wave falls on the plane boundary of the media interface in the open space. To get an obvious from a physical point of view solution of the problem of reflection and refraction of the wave, we should use integral representations of the solutions of Maxwell equations, or use the Fourier transforms of the unknown functions.

In the study of wave diffraction problems by bodies and screens, the method of integral equations is widely used [3], [4], [5]. Among the publications of recent years, we will mention the works [6], [8]. General theory of solvability of diffraction problems of electromagnetic waves by dielectric bodies with non-coordinate boundaries and conductive screens of intricate shape (the existence and uniqueness theorems, statements on the smoothness of the solution, etc.) is constructed using a complex mathematical apparatus, i.e., the method of pseudo-differential operators [9], [10].

For the approximate solution of diffraction problems by bodies and screens, various methods are used, both heuristic ([2], Ch. II) and strict ([2], Ch. III). Algorithms for numerical solution of integral equations of diffraction problems, especially in three-dimensional case, require significant computing resources. Perhaps for this reason, the method of the volume integral equation is not widely used [11], [12]. In waveguide theory, the [13] joining method is often used, by which diffraction problems are reduced directly to infinite sets of linear algebraic equations (ISLAE). In this case, the unknown quantities are coefficients of expansion of the searched fields in terms of the eigen waves of coordinate regions. The components of eigen waves can be found by the method of full or partial separation of variables.

In this chapter, we explore the possibility of numerical solving diffraction problems by bodies with non-coordinate boundaries, including the presence of thin conductive screens on the boundary, by the joining method. At first, the problems of scattering of electromagnetic wave by a non-coordinate media interface and by a conductive thin screen in a plane waveguide are considered. Exactly the same technique is used in the solving a two-dimensional and a three-dimensional problems of diffraction of a plane wave by a periodically perturbed media interface in the open space. Expansion of the unknown field in terms of cylindrical and spherical waves are used in reducing the problem of diffraction of an electromagnetic wave by a cylindrical rod with a non-coordinate boundary and wave diffraction problems by an axial-symmetrical body to ISLAE. Some of these problems were discussed earlier in the works [14]–[17].

## 2. Non-Coordinate Media Interface in a Plane Waveguide

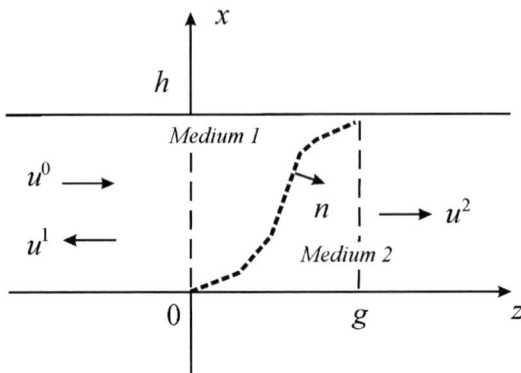

Figure 1. Media interface in a plane waveguide.

### 2.1. Formulation of the Scattering Problem

If the field depends on the time as $e^{-i\omega t}$, non-zero components of electric and magnetic vector $\mathbf{E} = (0, E_y, 0)$ and $\mathbf{H} = (H_x, 0, H_z)$ in the case of TE-waves have the form

$$E_y = u, \quad H_x = \frac{-1}{i\omega\mu}\frac{\partial u}{\partial z}, \quad H_z = \frac{1}{i\omega\mu}\frac{\partial u}{\partial x}, \tag{1}$$

where $u(x, z)$ is a solution of the Helmholtz equation.

Let the function $z = f(x) \geq 0$, $0 \leq x \leq h$, set the media interface. If the normal vector to the boundary has the form $\mathbf{n}(x) = (n_x(x), 0, n_z(x))$, then we have

$$[\mathbf{n}, \mathbf{E}] = (-n_z E_y, 0, n_x E_y), \quad [\mathbf{n}, \mathbf{H}] = (0, n_z H_x - n_x H_z, 0).$$

Therefore, the tangent components of the vectors $\mathbf{E}$ and $\mathbf{H}$ are continuous on the media interface if and only if the functions $u$ and $n_x \dfrac{\partial u}{\partial x} + n_z \dfrac{\partial u}{\partial z}$ are continuous.

Denote

$$\varphi_n(x) = \sqrt{2/h} \sin\frac{\pi n}{h}x, \quad n = 1, 2, \ldots$$

Let
$$u^0(x,z) = a_l^0 e^{i\gamma_l^1 z} \varphi_l(x)$$
be a potential function of the wave running onto the media interface from the left (see Figure 1). Let's look for potential functions of the wave reflected to the left and the wave that passed to the right in the form

$$u^1(x,z) = \sum_{n=1}^{+\infty} a_n e^{-i\gamma_n^1 (z-g)} \varphi_n(x), \quad u^2(x,z) = \sum_{n=1}^{+\infty} b_n e^{i\gamma_n^2 z} \varphi_n(x). \tag{2}$$

Here the longitudinal propagation constants $\gamma_n^j = \sqrt{k_j^2 - (\pi n/h)^2}$ are either real positive numbers, or imaginary numbers with a positive imaginary part.

Shift of the argument $z$ by $g = \max_{0 \le x \le h} f(x)$ is necessary for the following reasons. If $\gamma_n^1$ are real numbers, then exponents in the expression $u^1(x,z)$ are limited for any $z$. If $\gamma_n^1$ are imaginary numbers, then for positive $z$ exponents $e^{-i\gamma_n^1 z}$ become too large with increasing $n$. If the value of $g$ is not very large and when calculating the values of $n$ are also small, then we may not use such a shift.

The conjugation conditions are reduced to two series functional equations (SFE)

$$a_l^0 \psi_l^0(x) + \sum_{n=1}^{+\infty} a_n \psi_n^1(x) = \sum_{n=1}^{+\infty} b_n \psi_n^2(x), \quad a_l^0 \chi_l^0(x) + \sum_{n=1}^{+\infty} a_n \chi_n^1(x) = \sum_{n=1}^{+\infty} b_n \chi_n^2(x), \tag{3}$$

where $x \in [0, h]$ and

$$\psi_l^0(x) = e^{i\gamma_l^1 f(x)} \varphi_l(x), \quad \psi_n^1(x) = e^{i\gamma_n^1 (g-f(x))} \varphi_n(x), \quad \psi_n^2(x) = e^{i\gamma_n^2 f(x)} \varphi_n(x),$$

$$\chi_l^0(x) = e^{i\gamma_l^1 f(x)} \left[ n_x(x) \varphi_l'(x) + n_z(x) i\gamma_l^1 \varphi_l(x) \right],$$

$$\chi_n^1(x) = e^{i\gamma_n^1 (g-f(x))} \left[ n_x(x) \varphi_n'(x) - n_z(x) i\gamma_n^1 \varphi_n(x) \right],$$

$$\chi_n^2(x) = e^{i\gamma_n^2 f(x)} \left[ n_x(x) \varphi_n'(x) + n_z(x) i\gamma_n^2 \varphi_n(x) \right].$$

To proceed to ISLAE, let's project SFE on the functions $\varphi_m(x)$. Then we get ISLAE, consisting of two groups of equations:

$$a_l^0 \psi_{lm}^0 + \sum_{n=1}^{+\infty} a_n \psi_{nm}^1 = \sum_{n=1}^{+\infty} b_n \psi_{nm}^2, \quad a_l^0 \chi_{lm}^0 + \sum_{n=1}^{+\infty} a_n \chi_{nm}^1 = \sum_{n=1}^{+\infty} b_n \chi_{nm}^2, \quad (4)$$

where

$$\psi_{lm}^0 = \int_0^h \psi_l^0(x)\,\varphi_m(x)\,dx, \quad \chi_{lm}^0 = \int_0^h \chi_l^0(x)\,\varphi_m(x)\,dx,$$

$$\psi_{nm}^1 = \int_0^h \psi_n^1(x)\,\varphi_m(x)\,dx, \quad \chi_{nm}^1 = \int_0^h \chi_n^1(x)\,\varphi_m(x)\,dx, \quad (5)$$

$$\psi_{nm}^2 = \int_0^h \psi_n^2(x)\,\varphi_m(x)\,dx, \quad \chi_{nm}^2 = \int_0^h \chi_n^2(x)\,\varphi_m(x)\,dx.$$

## 2.2. A Particular Case: Inclined Boundary

Let $f(x) = x\,\mathrm{ctg}\,\alpha$, where $\mathrm{ctg}\,\alpha = g/h$ (a rectilinear inclined media interface). Then $n_x = -\cos\alpha$, $n_z = \sin\alpha$.

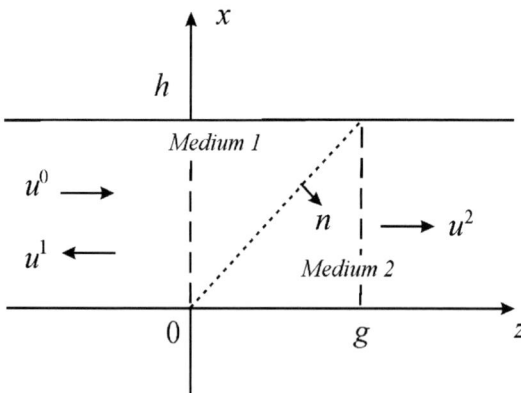

Figure 2. Inclined media interface

In this particular case, all integrals included in the ISLAE coefficients can be calculated analytically. All these integrals are expressed in terms of integrals of the form

$$I_{ss}(a,b,c) = \int_0^h e^{ax} \sin bx \cdot \sin cx \cdot dx, \quad I_{sc}(a,b,c) = \int_0^h e^{ax} \sin bx \cdot \cos cx \cdot dx.$$

The parameter $a$ can be either a real number or a complex number. Note that if the values $n$ and $r$ are large, then we have integrals of fast oscillating functions.

In the general case, it is expedient to approximate the media interface by a broken line with nodes at the points $(x_0, z_0) = (0,0), (x_1, z_1), \ldots (x_n, z_n) = (h, g)$. Then each individual integral in the sum

$$\int_0^h f(x)\, dx = \sum_{j=0}^{n-1} \int_{x_j}^{x_{j+1}} f(x)\, dx$$

can be calculated analytically also.

## 2.3. Exclusion of Some Unknowns

We can exclude unknowns $b_n$ from ISLAE (4), for example.

Let's move on to the vector-matrix form of equations. Let $a$, $b$ be the unknown infinite-dimensional vectors, $\psi_l^0$, $\chi_l^0$ and $\Psi^1$, $\Psi^2$, $X^1$, $X^2$ be known vectors and matrices. We write down ISLAE (4) in the form

$$a_l^0 \psi_l^0 + a\,\Psi^1 = b\,\Psi^2, \quad a_l^0 \chi_l^0 + a\,X^1 = b\,X^2,$$

it is convenient here and further to multiply vectors by matrices on the right. From the second equation, we have

$$b = a_l^0 \chi_l^0 (X^2)^{-1} + a\,X^1 (X^2)^{-1}. \tag{6}$$

Denote $K = (X^2)^{-1}\Psi^2$. Then

$$a_l^0 \left[\psi_l^0 - \chi_l^0 K\right] + a\left[\Psi^1 - X^1 K\right] = 0$$

or

$$a_l^0 \left[\psi_{lm}^0 - \sum_{r=1}^{+\infty} \chi_{lr}^0 \varepsilon_{rm}\right] + \sum_{n=1}^{+\infty} a_n \left[\psi_{nm}^1 - \sum_{r=1}^{+\infty} \chi_{nr}^1 \varepsilon_{rm}\right] = 0, \quad m = 1, 2, \ldots \tag{7}$$

or

$$\sum_{n=1}^{+\infty} a_n p_{nm} = -a_l^0 q_m, \quad m = 1, 2, \ldots,$$

$$p_{mn} = \psi_{nm}^1 - \chi_{nr}^1 \varepsilon_{rm}, \quad q_m = \psi_{lm}^0 - \sum_{r=1}^{+\infty} \chi_{lr}^0 k a_{rm}.$$

The elements of the matrix $K$ are such numbers $\varepsilon_{rm}$ that

$$\sum_{r=1}^{+\infty} \chi_{nr}^2 \varepsilon_{rm} = \psi_{nm}^2, \quad n, m = 1, 2, \ldots \qquad (8)$$

Here, the values $\chi_{nr}^2$ are independent of $m$, so equations (7) can be considered as an infinite set of infinite sets of linear algebraic equations (this can be used for parallelizing an algorithm).

After the values $a_n$ are found, the coefficients $b_n$ can be found by the formula (6) or, more simply, from the equations of the first group of ISLAE (4):

$$\sum_{n=1}^{+\infty} b_n \psi_{nm}^2 = \sum_{n=1}^{+\infty} a_n \psi_{nm}^1 + a_l^0 \psi_{lm}^0, \quad n = 1, 2, \ldots \qquad (9)$$

In exactly the same way, the unknowns $a_n$ could be excluded from the ISLAE.

We will point out two more variants of ISLAE for unknown coefficients:

$$\left[ a_l^0 \, \psi_l^0 + a \, \Psi^1 \right] (\Psi^2)^{-1} = \left[ a_l^0 \, \chi_l^0 + a \, X^1 \right] (X^2)^{-1},$$

$$\left[ b \, \Psi^2 - a_l^0 \, \psi_l^0 \right] (\Psi^1)^{-1} = \left[ b \, X^2 - a_l^0 \, \chi_l^0 \right] (X^1)^{-1}.$$

### 2.4. Computational Experiment

The approximate solution of all ISLAE can be found by the truncation method. The integrals (5) should be calculated preliminary.

In the case of the set of equations (4), we will leave $N$ unknown $a_n$ and $b_n$, and $N$ equations from each part of the ISLAE. We can also use four truncation parameters $N_1$, $N_2$, $M_1$, $M_2$ respectively, if only the condition $N_1 + N_2 = M_1 + M_2$ was fulfilled.

When truncating ISLAE (7), we proceed as follows. Replace ISLAE with finite set of linear algebraic equations (SLAE)

$$\sum_{n=1}^{N} P_{kn}\, a_n = -Q_k\, a_l^0, \quad k = 1..N,$$

here $N$ is the truncation parameter. Then the integrals $\psi_{lk}^0$ should be calculated for a fixed $l$ and for $k = 1..N$, and the integrals $\psi_{nk}^1$ should be calculated for $n = 1..N$, $k = 1..N$. Let's introduce another truncation parameter $R \geq N$. Then $\chi_{lr}^0$ are calculated for the fixed $l$ and for $r = 1..R$, and $\chi_{nr}^1$ is calculated when $n = 1..N$, $r = 1..R$.

The numbers $\varepsilon_{rk}$ will be needed when $r = 1..R$, $k = 1..N$. ISLAE (8) after truncation has the form

$$\sum_{r=1}^{R} \chi_{nr}^2\, K_{rm} = \psi_{nm}^2, \quad n = 1..R, \quad m = 1..N.$$

Therefore, the values of the integrals $\chi_{nr}^2$ and $\psi_{nm}^2$, $n = 1..R$, $r = 1..R$, $m = 1..N$ should be found preliminary.

Finally, ISLAE (9) is truncated to SLAE

$$\sum_{n=1}^{N} \psi_{nm}^2\, b_n = \sum_{n=1}^{N} \psi_{mn}^1\, a_n + a_l^0\, \psi_{lm}^0, \quad m = 1..N.$$

The computational experiment has shown that with the correct numerical solving of ISLAE (4) and (7)-(9) we obtain the same values of the coefficients $a_n$ and $b_n$. The correctness and accuracy of the numerical solution can be evaluated using the energy conservation law

$$\sum_{n=1}^{+\infty} |a_n|\, \mathrm{Re}\, \gamma_n^1 - \sum_{n=1}^{+\infty} |b_n|\, \mathrm{Re}\, \gamma_n^2 = |a^0|\, \mathrm{Re}\, \gamma_l^1,$$

here only those terms remain in the sums in which $\mathrm{Re}\, \gamma_n^j \neq 0$.

In the case of an inclined media interface, reliable results were obtained with some restrictions on the values of the angle $\alpha$ (depending on the values of dielectric constants $\varepsilon_1$ and $\varepsilon_2$) and therefore, on the values of $g$. In the general case, the values of $g$ also cannot be too large. The most accurate solution is found when the function $f(x)$ is monotonous.

## 3. Curvilinear Screen in a Plane Waveguide

What will change if an ideally conductive infinitely thin plate (screen) $\mathcal{M}$ is placed on the media interface?

### 3.1. Paired Series Functional Equation

Suppose for simplicity of reasoning that the waveguide is filled with a homogeneous medium, the longitudinal propagation constants $\gamma_n^1 = \gamma_n^2 = \gamma_n$ and therefore the screen $\mathcal{M}$ is placed on the conditional media interface $z = f(x)$. Let also the wave from the external source be given to the left and to the right of the screen. Then, as it is shown in [14], the conjugation conditions

$$\sum_{n=1}^{+\infty} a_n \psi_n^1(x) = \sum_{n=1}^{+\infty} b_n \psi_n^2(x), \quad \sum_{n=1}^{+\infty} a_n \chi_n^1(x) = \sum_{n=1}^{+\infty} b_n \chi_n^2(x), \quad (10)$$

should be fulfilled on the complement $\mathcal{N}$ of the screen to the whole media interface, and boundary conditions

$$a_l^0 \psi_l^0(x) + \sum_{n=1}^{+\infty} a_n \psi_n^1(x) = 0, \quad a_l^0 \psi_l^0(x) + \sum_{n=1}^{+\infty} b_n \psi_n^2(x) = 0 \quad (11)$$

should be fulfilled on $\mathcal{M}$.

Therefore, the first condition in (10) should be fulfilled both on $\mathcal{N}$ and on $\mathcal{M}$. Let's introduce new unknowns $c_n$ so that

$$\sum_{n=1}^{+\infty} c_n \varphi_n(x) = \sum_{n=1}^{+\infty} a_n \psi_n^1(x) = \sum_{n=1}^{+\infty} b_n \psi_n^2(x). \quad (12)$$

We project equalities (12) on the functions $\varphi_m(x)$ and get

$$\sum_{n=1}^{+\infty} a_n \psi_{nm}^1 = c_m, \quad \sum_{n=1}^{+\infty} b_n \psi_{nm}^2 = c_m \quad \text{or} \quad a \Psi^1 = c, \quad b \Psi^2 = c.$$

Let $P = (\Psi^1)^{-1}$, $Q = (\Psi^2)^{-1}$. Then

$$a = c P, \quad b = c Q \quad \text{or} \quad a_n = \sum_{m=1}^{+\infty} c_m p_{mn}, \quad b_n = \sum_{m=1}^{+\infty} c_m q_{mn}.$$

Thus, the conditions (2.1) and (2.2) are converted to paired series functional equation (PSFE) on the segment $[0, h]$. Its first part on $\mathcal{M}$ has the form

$$\sum_{n=1}^{+\infty} c_n \, \varphi_n(x) = -a_l^0 \, \psi_l^0(x), \tag{13}$$

and the second part of the PSFE on $\mathcal{N}$ is derived from the second equality in the formula (10):

$$\sum_{m=1}^{+\infty} c_m \, w_m(x) = 0, \quad w_m(x) = \sum_{n=1}^{+\infty} \left[ p_{mn} \, \chi_n^1(x) - q_{mn} \, \chi_n^2(x) \right] \tag{14}$$

or, in vector-matrix form, $c \, w(x) = 0$ on $\mathcal{N}$, where the vector-function $w(x) = P \, \chi^1(x) + Q \, \chi^2(x)$.

## 3.2. Infinite Set of Linear Algebraic Equations

If we project PSFE (13), (14) on the functions $\varphi_k(x)$, then we get the ISLAE with an ill conditioned matrix of coefficients. To move on to regular ISLAE, let's build an integral transformation with the kernel $L(t, x)$, which converts the functions $w_m(x)$ into the functions $\varphi_m(x)$:

$$\int_0^h w_m(t) \, L(t, x) \, dt = \varphi_m(x), \quad x \in [0, h], \quad m = 1, 2, \ldots \tag{15}$$

The kernel of such transformation will be searched in the form

$$L(t, x) = \sum_{r=1}^{+\infty} \sum_{s=1}^{+\infty} L_{rs} \, \varphi_r(t) \, \varphi_s(x).$$

It is easy to see that the equations (15) will be fulfilled if

$$\sum_{r=1}^{+\infty} w_{mr} L_{rs} = \delta_{ms}, \quad m = 1, 2, \ldots$$

($\delta_{ms}$ is Kronecker symbol), where

$$w_{mr} = \int_0^h w_m(t) \, \varphi_r(t) \, dt = \sum_{n=1}^{+\infty} [p_{mn} \, X_{nr}^1 - q_{mn} \, X_{nr}^2].$$

In other words, the matrix $L = \{L_{rs}\}$ should be the inverse matrix to the matrix $\Omega = \{\omega_{mr}\} = [\Psi^1]^{-1} X^1 - [\Psi^2]^{-1} X^2$.

Hence, there is an integral-series identity (ISI)

$$\int_0^h \left(\sum_{m=1}^{+\infty} c_m \omega_m(t)\right) L(t, x)\, dt = \sum_{m=1}^{+\infty} c_m \varphi_m(x), \quad x \in [0, h] \quad (16)$$

(for any set of numbers $c_m$ at which the series converge). Then we have on $\mathcal{N}$

$$\sum_{m=1}^{+\infty} c_m \varphi_m(x) = \sum_{m=1}^{+\infty} c_m \sum_{r=1}^{+\infty} \sum_{s=1}^{+\infty} L_{rs} \Omega_{mr}^{\mathcal{M}} \varphi_s(x), \quad (17)$$

where

$$\Omega_{mr}^{\mathcal{M}} = \int_{\mathcal{M}} \omega_m(t)\, \varphi_r(t)\, dt = \sum_{n=1}^{+\infty} [p_{mn} X_{nr}^{1,\mathcal{M}} - q_{mn} X_{nr}^{1,\mathcal{M}}],$$

$$X_{nr}^{1,\mathcal{M}} = \int_{\mathcal{M}} \chi_n^1(t)\, \varphi_r(t)\, dt, \quad X_{nr}^{2,\mathcal{M}} = \int_{\mathcal{M}} \chi_n^2(t)\, \varphi_r(t)\, dt.$$

Let's project PSFE (13), (17) on the functions $\varphi_k(x)$, $k = 1, 2, \ldots$ and we will finally get ISLAE

$$c_k = \sum_{m=1}^{+\infty} c_m \sum_{r=1}^{+\infty} \sum_{s=1}^{+\infty} L_{rs} \Omega_{mr}^{\mathcal{M}} \Phi_{sk}^{\mathcal{N}} - a_l^0 \Psi_{lk}^{0,\mathcal{M}}, \quad k = 1, 2, \ldots, \quad (18)$$

where

$$\Phi_{sk}^{\mathcal{N}} = \int_{\mathcal{N}} \varphi_s(x)\, \varphi_k(x)\, dx = \delta_{sk} - \Phi_{sk}^{\mathcal{M}}, \quad \Psi_{lk}^{0,\mathcal{M}} = \int_{\mathcal{M}} \psi_l^0(x)\, \varphi_k(x)\, dx.$$

### 3.3. Particular Case: Lateral Screen

Let the thin conductive screen be placed on the cross-section of the plane waveguide. In this case, $f(x) \equiv 0$ and $g = 0$. Then $n_x = 0$, $n_z = 1$,

$$\psi_l^0(x) = \varphi_l(x), \quad \psi_n^1(x) = \psi_n^2(x) = \varphi_n(x),$$

$$\chi_l^0(x) = i\gamma_l\,\varphi_l(x), \quad -\chi_n^1(x) = \chi_n^2(x) = i\gamma_n\,\varphi_n(x)$$

and therefore

$$\psi_{lm}^0 = \delta_{lm},\ \psi_{nm}^1 = \psi_{nm}^2 = \delta_{nm},\ \chi_{lm}^0 = i\gamma_l\,\delta_{lm},\ -\chi_{nm}^1 = \chi_{nm}^2 = i\gamma_n\,\delta_{nm}.$$

From (12) we have $c_n = a_n = b_n$. Further, $p_{mn} = q_{mn} = \delta_{mn}$ and $\omega_m(x) = -2i\gamma_m\,\varphi_m(x)$, $\omega_{mr} = -2i\gamma_m\,\delta_{mr}$. The matrix $L$ becomes diagonal, $L_{rs} = \dfrac{i}{2\gamma_r}\,\delta_{rs}$ and

$$L(t,x) = \frac{i}{2} \sum_{r=1}^{+\infty} \frac{1}{\gamma_r}\,\varphi_r(t)\,\varphi_r(x).$$

Finally, ISLAE (18) is also significantly simplified:

$$c_k = \sum_{m=1}^{+\infty} c_m\,\gamma_m \sum_{r=1}^{+\infty} \frac{1}{\gamma_r}\,I_{mr}\,J_{rk} - a_r^0\,I_{lk}, \quad k = 1, 2, \ldots,$$

where

$$I_{mr} = \int_M \varphi_m(x)\,\varphi_r(x)\,dx, \quad J_{rk} = \int_N \varphi_r(x)\,\varphi_k(x)\,dx = \delta_{rk} - I_{rk}.$$

## 4. Periodic Media Interface

The problem of scattering an electromagnetic wave by a non-coordinate periodic media interface in the open space is examined according to the same scheme as the problem on the fall of a wave on a non-coordinate media interface in a plane waveguide.

### 4.1. Two-Dimensional Periodic Problem

Let's consider a two-dimensional special case of a general problem when the electromagnetic field does not depend on the spatial coordinate $y$. Let's limit ourselves to the case of parallel polarization of the field when its non-zero components have the form (1).

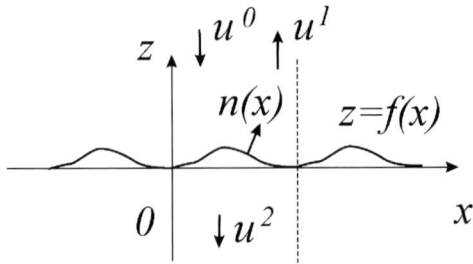

Figure 3. Periodic media interface.

Let the media interface be given by the function $z = f(x) \geq 0$, where $f(x)$ is a $p$-periodic function, the normal vector to the boundary is $n = (n_x(x), 0, n_z(x))$. Let's denote $h = \max_{x \in [0,p]} f(x)$. Suppose that the magnetic permeability of the media is the same, but the dielectric constants, and hence the wave numbers, are different.

Let the plane wave falling from above (see Figure 3) have a potential function
$$u^0(x, z) = a^0 \, e^{-ik_1 \sin \theta^0 \cdot x - ik_1 \cos \theta^0 \cdot z}.$$

Let's look for potential functions of the reflected wave and the refracted wave in the form

$$u^1(x,z) = e^{i\alpha x} \sum_{n=-\infty}^{+\infty} a_n \, e^{i\gamma_n^1 z} \, e^{i\frac{2\pi}{p} n x}, \quad u^2(x,z) = e^{i\alpha x} \sum_{n=-\infty}^{+\infty} b_n \, e^{-i\gamma_n^2 (z-h)} \, e^{i\frac{2\pi}{p} n x}. \tag{19}$$

The values of constants $\gamma_n^{1,2} = \sqrt{k_{1,2}^2 - (\alpha + 2\pi n/p)^2}$ are either positive real numbers, or imaginary numbers with a positive imaginary part. Shift by the value $h$ limits exponential growth for $n \to \pm\infty$.

As in the case of TE-waves in a plane waveguide, the conjugation conditions of the field on the media interface are equivalent to the continuity condition of functions $u$ and $n_x \dfrac{\partial u}{\partial x} + n_z \dfrac{\partial u}{\partial z}$ when $z = f(x)$. From the conjugation conditions it follows that the Floquet parameter $\alpha = -k_1 \sin \theta^0$. Then $k_1 \cos \theta^0 = \gamma_0^1$.

Let's denote $p_n = \alpha + 2\pi n/p$. Coefficients $a_n$ and $b_n$ of expansions of quasi-periodic functions should be found from the series functional equa-

tions
$$a^0\,\psi^0(x) + \sum_{n=-\infty}^{+\infty} a_n\,\psi_n^1(x) = \sum_{n=-\infty}^{+\infty} b_n\,\psi_n^2(x), \qquad (20)$$

$$a^0\,\chi^0(x) + \sum_{n=-\infty}^{+\infty} a_n\,\chi_n^1(x) = \sum_{n=-\infty}^{+\infty} b_n\,\chi_n^2(x), \qquad (21)$$

where
$$\psi^0(x) = e^{-i\gamma_0^1 f(x)}, \quad \psi_n^1(x) = e^{i\gamma_n^1 f(x)}\,e^{i\frac{2\pi}{p}nx}, \quad \psi_n^2(x) = e^{i\gamma_n^2 [h-f(x)]}\,e^{i\frac{2\pi}{p}nx},$$

$$\chi^0(x) = e^{-i\gamma_0^1 f(x)}\,[n_x(x)\,\alpha - n_z(x)\,\gamma_0^1],$$

$$\chi_n^1(x) = e^{i\gamma_n^1 f(x)}\,e^{i\frac{2\pi}{p}nx}\,[n_x(x)\,p_n + n_z(x)\,\gamma_n^1],$$

$$\chi_n^2(x) = e^{i\gamma_n^2 [h-f(x)]}\,e^{i\frac{2\pi}{p}nx}\,[n_x(x)\,p_n - n_z(x)\,\gamma_n^2].$$

Equality (20) should be fulfilled at all points of the segment $[0,p]$, and equality (21) should be fulfilled everywhere in the interval $(0,p)$, except for points where, perhaps, the smoothness of the line $z = f(x)$ is broken.

Let's project equalities (20) and (21) on the function $e^{i\frac{2\pi}{p}kx}$, that is, multiply them by $e^{-i\frac{2\pi}{p}kx}$ and integrate from 0 to $p$. Then we get ISLAE

$$a^0\,\psi_k^0 + \sum_{n=-\infty}^{+\infty} a_n\,\psi_{nk}^1 = \sum_{n=-\infty}^{+\infty} b_n\,\psi_{nk}^2, \qquad (22)$$

$$a^0\,\chi_k^0 + \sum_{n=-\infty}^{+\infty} a_n\,\chi_{nk}^1 = \sum_{n=-\infty}^{+\infty} b_n\,\chi_{nk}^2, \qquad (23)$$

where
$$\psi_k^0 = \int_0^p \psi^0(x)\,e^{-i\frac{2\pi}{p}kx}\,dx, \quad \chi_k^0 = \int_0^p \chi^0(x)\,e^{-i\frac{2\pi}{p}kx}\,dx,$$

$$\psi_{nk}^1 = \int_0^p \psi_n^1(x)\,e^{-i\frac{2\pi}{p}kx}\,dx, \quad \chi_{nk}^1 = \int_0^p \chi_n^1(x)\,e^{-i\frac{2\pi}{p}kx}\,dx, \qquad (24)$$

$$\psi_{nk}^2 = \int_0^p \psi_n^2(x)\,e^{-i\frac{2\pi}{p}kx}\,dx, \quad \chi_{nk}^2 = \int_0^p \chi_n^2(x)\,e^{-i\frac{2\pi}{p}kx}\,dx.$$

As in the case of a plane waveguide, it is possible to exclude unknowns $a_n$ from equations (22), (23). Let's write down these equations in vector-matrix form (for infinite-dimensional vectors and matrices)

$$\psi^0 + a\,\Psi^1 = b\,\Psi^2, \quad \chi^0 + a\,X^1 = b\,X^2.$$

Here, it is also convenient to use multiplication by the matrices on the right.

Let the matrix $K$ be such that $X^2 K = \Psi^2$. Multiply the second equality by $K$ on the right and subtract the first equality from here. We get

$$a\,(X^1 K - \Psi^1) = \psi^0 - \chi^0 K$$

or

$$\sum_{n=-\infty}^{+\infty} a_n \left[ \sum_{r=-\infty}^{+\infty} K_{rk}\chi^1_{nr} - \psi^1_{nk} \right] = \psi^0_k - \sum_{r=-\infty}^{+\infty} K_{rk}\chi^0_r, \quad k = 0, \pm 1, \ldots \tag{25}$$

Unknowns $b_n$ can be found by the formula $b = \left[\psi^0 + a\,\Psi^1\right](\Psi^2)^{-1}$.

Suppose that the media interface is a broken line. Let

$$f(x) = \left\{ 0 \leq x \leq q : \frac{h}{q}x; \quad q \leq x \leq p : \frac{h}{p-q}(p-x) \right\},$$

two parameters $q$ and $h$ determine the shape of the boundary. In this case, we can get explicit expressions for all integrals (24).

## 4.2. Computational Experiment

A computational experiment has shown that during a numerical solving the truncated ISLAE (23), (24) and (25), we obtain almost identical values of the coefficients $a_n$ and $b_n$. During test calculations, the accuracy of the calculations deteriorates with the increase in the values of the parameter $h$.

For $h = 0$, as we expected, only one of the coefficients $a_n$ and only one of the coefficients $b_n$ are not zero. Their values coincide with the exact solution of the classical problem of reflection and refraction of a plane wave on the plane media interface.

The following effect was detected. Values of coefficients $a_n$ and $b_n$ with relatively small numbers are found with good accuracy both for low values of the truncation parameter $N$ and for high values of this parameter. But as $N$ increases, the values $|a_n|$ and $|b_n|$ for the extreme values $n$ become very large. This is explained by the fact that the series (19) should not converge for $z = 0$. At the same time, for $z \neq 0$ exponential multipliers in the sums compensate the growth of the coefficients $a_n$ and $b_n$ when $n \to \pm\infty$. Therefore, it is expedient to understand the solutions of such diffraction problems in a generalized sense [19], [20].

It is also possible to move on from SFE (20), (21) to ISLAE by the collocation method, that is, to write down equations (20) and (21) in the finite number of points $x_j \in [0, p]$:

$$\sum_{n=-N}^{N} a_n \psi_n^1(x_j) - \sum_{n=-N}^{N} b_n \psi_n^2(x_j) = -a^0 \psi^0(x_j),$$

$$\sum_{n=-N}^{N} a_n \chi_n^1(x_j) - \sum_{n=-N}^{N} b_n \chi_n^2(x_j) = -a^0 \chi^0(x_j).$$

In the simplest case, we have $x_j = pj/(2N+1)$, $j = 0..2N$. When $h = 0$, the approximate solution of this ISLAE is exactly the same as in the case of ISLAE (23), (24), but with increasing $h$, the accuracy decreases much faster.

## 4.3. Two-Periodic Non-Coordinate Boundary

Let in the three-dimensional case the media interface be given by equation $z = f(x, y)$, then the normal at the boundary point is $(-f'_x, -f'_y, 1)$, where $f'_x = \partial f/\partial x$, $f'_y = \partial f/\partial y$. The local basis in the tangential plane is formed by vectors $(1, 0, f'_x)$ and $(0, 1, f'_y)$. As it is shown in [15], if $z = f(x, y)$ then the expressions $(1 + f'^2_y) E_x - f_x f_y E_y + f_x E_z$, $-f_x f_y E_x + (1 + f'^2_x) E_y + f_y E_z$, $(1 + f'^2_y) H_x - f_x f_y H_y + f_x H_z$ and $-f_x f_y H_x + (1 + f'^2_x) H_y + f_y H_z$ should be continuous.

Let a plane electromagnetic wave run over the media interface. If the function $f(x, y)$ is two-periodic with periods of $p$ and $q$ by variables $x$ and $y$ respectively, then all components of the electromagnetic field can be searched in the form of sums of Floquet series of the form

$$\sum_{m=-\infty}^{+\infty} \sum_{n=-\infty}^{+\infty} a_{m,n}(z) e^{ip_m x + iq_n y},$$

where $p_m = \alpha + \frac{2\pi}{p}m$, $q_n = \beta + \frac{2\pi}{q}n$. From Maxwell equations it follows that for each value $m$ and $n$ the Floquet coefficients of the field components should be solutions of a set of differential equations

$$iq_n H_{z,mn} - H'_{y,mn} = -i\omega\varepsilon_0\varepsilon E_{x,mn}, \quad iq_n E_{z,mn} - E'_{y,mn} = i\omega\mu_0\mu H_{x,mn},$$

$$H'_{x,mn} - ip_m H_{z,mn} = -i\omega\varepsilon_0\varepsilon E_{y,mn}, \quad E'_{x,mn} - ip_m E_{z,mn} = i\omega\mu_0\mu H_{y,mn},$$

$$ip_m H_{y,mn} - iq_n H_{x,mn} = -i\omega\varepsilon_0\varepsilon E_{z,mn}, \quad ip_m E_{y,mn} - iq_n E_{x,mn} = i\omega\mu_0\mu H_{z,mn}.$$

Let $\gamma_{mn} = \sqrt{k^2 - p_m^2 - q_n^2}$ and these numbers are either real positive numbers, or imaginary numbers with a positive imaginary part. Let's select $E_x$ and $E_y$ as potential functions. Then (see [15])

$$E_{x,mn} = a_{m,n}^+ e^{i\gamma_{m,n}z} + a_{m,n}^- e^{-i\gamma_{m,n}z}, \quad E_{y,mn} = b_{m,n}^+ e^{i\gamma_{m,n}z} + b_{m,n}^- e^{-i\gamma_{m,n}z},$$

where $a_{mn}^\pm$, $b_{mn}^\pm$ are some constants. When $a_{mn}^- = 0$, $b_{mn}^- = 0$ we have waves of positive orientation, and when $a_{mn}^+ = 0$, $b_{mn}^+ = 0$ we have waves of negative orientation. All other field components are expressed through functions $E_x$ and $E_y$.

We will look for potential functions of the wave, outgoing into the region $z > f(x,y)$, in the form

$$E_x^1 = \sum_{m=-\infty}^{+\infty} \sum_{n=-\infty}^{+\infty} a_{mn}^1 \gamma_{mn}^1 e^{i\gamma_{mn}^1 z} e^{ip_m x + iq_n y},$$

$$E_y^1 = \sum_{m=-\infty}^{+\infty} \sum_{n=-\infty}^{+\infty} b_{mn}^1 \gamma_{mn}^1 e^{i\gamma_{mn}^1 z} e^{ip_m x + iq_n y},$$

and we will look for the potential functions of the wave, outgoing into the region $z < f(x,y)$, in the form

$$E_x^2 = \sum_{m=-\infty}^{+\infty} \sum_{n=-\infty}^{+\infty} a_{mn}^2 \gamma_{mn}^2 e^{-i\gamma_{mn}^2 z} e^{ip_m x + iq_n y},$$

$$E_y^2 = \sum_{m=-\infty}^{+\infty} \sum_{n=-\infty}^{+\infty} b_{mn}^2 \gamma_{mn}^2 e^{-i\gamma_{mn}^2 z} e^{ip_m x + iq_n y}.$$

Here, the propagation constants $\gamma_{mn}^{1,2}$ are calculated by wave numbers $k_{1,2}$ of regions filled with the media with different properties.

Let the wave running from above on the media interface be the sum of two plane waves of different polarization and its potential wave functions have the form

$$E_x^0 = a^0 \, \gamma_{00}^1 \, e^{-i\gamma_{00}^1 z} \, e^{-ik_1 \sin\theta^0 \cdot x - ik_1 \cos\theta^0 \cdot y},$$

$$E_y^0 = b^0 \, \gamma_{00}^1 \, e^{-i\gamma_{00}^1 z} \, e^{-ik_1 \sin\theta^0 \cdot x - ik_1 \cos\theta^0 \cdot y}.$$

It follows from the conjugation conditions that the Floquet parameters have the form $\alpha = -k_1 \sin\theta^0$, $\beta = -k_1 \cos\theta^0$.

Thus, the problem of scattering a plane electromagnetic wave by a two-periodic media interface is reduced to four series functional equations to determine the coefficients $a_{mn}^1$, $b_{mn}^1$, $a_{mn}^2$, $b_{mn}^2$ of the form

$$a^0 \chi_0^j(x,y) + b^0 \psi_0^j(x,y) + \sum_{m=-\infty}^{+\infty}\sum_{n=-\infty}^{+\infty} [a_{mn}^1 \chi_{1,mn}^j(x,y) + b_{mn}^1 \psi_{1,mn}^j(x,y)] =$$

$$= \sum_{m=-\infty}^{+\infty}\sum_{n=-\infty}^{+\infty} [a_{mn}^2 \chi_{2,mn}^j(x,y) + b_{mn}^2 \psi_{2,mn}^j(x,y)], \quad j = 1,2,3,4. \quad (26)$$

For example, for the first conjugation condition ($j = 1$) we have

$$\chi_0^1(x,y) = \left[(1 + f_y^2(x,y))\gamma_{00}^1 + \alpha \, f_x(x,y)\right] e^{-i\gamma_{00}^1 f(x,y)},$$

$$\psi_0^1(x,y) = \left[-f_x(x,y)f_y(x,y)\gamma_{00}^1 + \beta \, f_x(x,y)\right] e^{-i\gamma_{00}^1 f(x,y)},$$

$$\chi_{1,mn}^1(x,y) = \left[(1 + f_y^2(x,y))\gamma_{mn}^1 - p_m \, f_x(x,y)\right] e^{i\gamma_{mn}^1 f(x,y)} e^{i\frac{2\pi}{p}mx + i\frac{2\pi}{q}ny},$$

$$\psi_{1,mn}^1(x,y) = \left[-f_x(x,y)f_y(x,y)\gamma_{mn}^1 - q_n \, f_x(x,y)\right] e^{i\gamma_{mn}^1 f(x,y)} e^{i\frac{2\pi}{p}mx + i\frac{2\pi}{q}ny},$$

$$\chi_{2,mn}^1(x,y) = \left[(1 + f_y^2(x,y))\gamma_{mn}^2 + p_m \, f_x(x,y)\right] e^{-i\gamma_{mn}^2 f(x,y)} e^{i\frac{2\pi}{p}mx + i\frac{2\pi}{q}ny},$$

$$\psi_{2,mn}^1(x,y) = \left[-f_x(x,y)f_y(x,y)\gamma_{mn}^2 + q_n \, f_x(x,y)\right] e^{-i\gamma_{mn}^2 f(x,y)} e^{i\frac{2\pi}{p}mx + i\frac{2\pi}{q}ny}.$$

Series functional equations are transformed into a large infinite set of linear algebraic equations, consisting of four groups of equations. An approximate solution of ISLAE can be found by truncation method. Some techniques for parallelizing the numerical solution algorithm of a truncated set of equations is described in [15].

# 5. Electromagnetic Wave Scattering by a Cylindrical Body

Let's consider the two-dimensional scattering problem of a plane parallel polarized electromagnetic wave by an infinite dielectric rod.

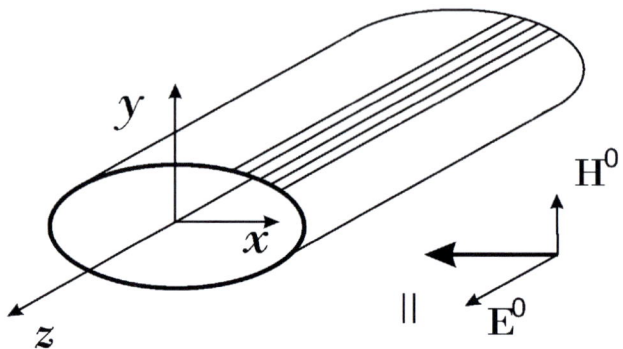

Figure 4. Plane wave runs over the dielectric body.

## 5.1. Paired Series Equations

It is convenient to use simultaneously Cartesian coordinates $(x, y, z)$, and cylindrical coordinates $(r, \alpha, z)$. Let the forming line of cylindrical boundary of the rod be parallel to the axis $z$, and the guiding line is a simple closed curve $r = f(\alpha)$, $0 \leq \alpha < 2\pi$, smooth or piecewise smooth.

If the electromagnetic field does not depend on the coordinate $z$, then the non-zero components of the waves of parallel polarization are expressed through the solution of the Helmholtz equation $u(r, \alpha)$ as follows:

$$E_z = u, \quad H_\alpha = \frac{-1}{i\omega\mu_0\mu} \frac{\partial u}{\partial r}, \quad H_r = \frac{1}{i\omega\mu_0\mu} \frac{1}{r} \frac{\partial u}{\partial \alpha}.$$

As it is shown in [16], the expressions $u$ and $\dfrac{\partial u}{\partial \alpha} \dfrac{f'(\alpha)}{f(\alpha)} - \dfrac{\partial u}{\partial r} f(\alpha)$ should be continuous on the media interface.

Suppose that a plane wave runs over a dielectric body. Its potential function we expand in terms of cylindrical waves:

$$E_z^0 = a^0 e^{ik_1 x} = a^0 \sum_{n=-\infty}^{+\infty} (-i)^n J_n(k_1 r) e^{in\alpha}.$$

Let's look for potential field functions outside the body and inside the body in the form

$$E_z^1 = \sum_{n=-\infty}^{+\infty} b_n H_n^{(1)}(k_1 r) e^{in\alpha}, \quad E_z^2 = \sum_{n=-\infty}^{+\infty} c_n J_n(k_2 r) e^{in\alpha}, \quad (27)$$

here $J_n(\cdot)$ and $H_n^{(1)}(\cdot)$ are Bessel functions and Hankel functions of the 1st kind. We assume that the magnetic permeabilities of the media are equal, $k_1$ and $k_2$ are real wave numbers.

Consequently, the problem on scattering a plane wave is reduced to two series functional equations

$$\sum_{n=-\infty}^{+\infty} b_n \varphi_n^I(\alpha) e^{in\alpha} - \sum_{n=-\infty}^{+\infty} c_n \chi_n^I(\alpha) e^{in\alpha} = -a^0 \sum_{n=-\infty}^{+\infty} \psi_n^I(\alpha) e^{in\alpha}, \quad (28)$$

$$\sum_{n=-\infty}^{+\infty} b_n \varphi_n^{II}(\alpha) e^{in\alpha} - \sum_{n=-\infty}^{+\infty} c_n \chi_n^{II}(\alpha) e^{in\alpha} = -a^0 \sum_{n=-\infty}^{+\infty} \psi_n^{II}(\alpha) e^{in\alpha}, \quad (29)$$

where

$$\varphi_n^I(\alpha) = H_n^{(1)}(k_1 f(\alpha)), \quad \chi_n^I(\alpha) = J_n(k_2 f(\alpha)), \quad \psi_n^I(\alpha) = (-i)^n J_n(k_1 f(\alpha)),$$

$$\varphi_n^{II}(\alpha) = in H_n^{(1)}(k_1 f(\alpha)) f'(\alpha)/f(\alpha) - k_1 H_n^{(1)'}(k_1 f(\alpha)) f(\alpha),$$

$$\chi_n^{II}(\alpha) = in J_n(k_2 f(\alpha)) f'(\alpha)/f(\alpha) - k_2 J_n'(k_2 f(\alpha)) f(\alpha),$$

$$\psi_n^{II}(\alpha) = (-i)^n \left[ in J_n(k_1 f(\alpha)) f'(\alpha)/f(\alpha) - k_1 J_n'(k_1 f(\alpha)) f(\alpha) \right].$$

In the particular case, when $f(\alpha) = R$ (we have a round rod), the conditions (28) and (29) become simple independent sets of linear equations

$$b_n H_n^{(1)}(k_1 R) - c_n J_n(k_2 R) = -a^0 (-i)^n J_n(k_1 R),$$

$$b_n H_n^{(1)'}(k_1 R) k_1 - c_n J_n'(k_2 R) k_2 = -a^0 (-i)^n J_n'(k_1 R) k_1.$$

Hence it is easy to find explicit expressions for the coefficients $b_n$ and $c_n$.

## 5.2. Computational Experiment

Let's move on from functional equations (28) and (29) to ISLAE. If we project equations (28) and (29) on the set of functions $e^{im\alpha}$, then we get a ISLA consisting of two groups of equations

$$\sum_{n=-\infty}^{+\infty} b_n \Phi^I_{n,n-m} - \sum_{n=-\infty}^{+\infty} c_n X^I_{n,n-m} = -a^0 \sum_{n=-\infty}^{+\infty} \Psi^I_{n,n-m}, \quad m = 0, \pm 1, \ldots \tag{30}$$

$$\sum_{n=-\infty}^{+\infty} b_n \Phi^{II}_{n,n-m} - \sum_{n=-\infty}^{+\infty} c_n X^{II}_{n,n-m} = -a^0 \sum_{n=-\infty}^{+\infty} \Psi^{II}_{n,n-m}, \quad m = 0, \pm 1, \ldots \tag{31}$$

Here we denote

$$F_{n,k} = \int_0^{2\pi} f_n(\alpha) \, e^{ik\alpha} \, d\alpha.$$

The calculation of such integrals in the general case requires considerable effort, because for high values $|k|$ the integrands are fast oscillating functions. The approximate solution of ISLAE (30), (31) is found by the truncation method. Both the coefficients $a_n$ and the coefficients of $b_n$ can be considered as unknowns. If we exclude the half of the unknowns, we will have a set of linear equations of less dimension.

We can also write down equations (30), (31) for $\alpha = \alpha_j$, where $\alpha_j$ is a finite set of the segment $[0, 2\pi]$ points (the collocation method). In this case, the amount of work in calculating the SLAE coefficients is significantly reduced. But as the computational experiment has shown, such an approach makes sense only when the shape of the cross-section of dielectric rod is close to a circle. Otherwise, computational error increases dramatically.

Figure 5 shows the dependence of the energy flow density on the variable $\alpha$ in the case when the conducting cylinder and its boundary are determined by the function of the form $f(\alpha) = R + p(\sin q\alpha - 1)$ for $p = 0.2$ and $q = 5, 10, 15$. When the cylinder boundary is wave distorted, the following effect is observed: the scattering diagram becomes smoother.

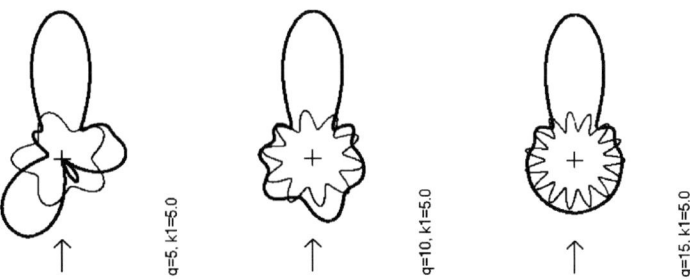

Figure 5. Scattering diagram for conductive cylinder with non-coordinate boundary.

## 5.3. Diffraction by a Thin Conductive Tape in a Homogeneous Space

Let the plane electromagnetic wave now run over the thin conductive strip, which is a part of a circular cylindrical surface $f(\alpha) = R$. The part $\mathcal{M}$ of the interval $[0, 2\pi)$ corresponds to the strip, and $\mathcal{N}$ is the rest part of this interval. Suppose, for simplicity of reasoning, that space is filled with a homogeneous medium and the plane wave is given for both regions $r > R$ and $r < R$. Potential functions of the scattered field will be searched in the form (27). Since the boundary conditions on $\mathcal{M}$ and the conjugation conditions on $\mathcal{N}$ have the form

$$E_z^0 + E_z^1 = 0, \quad E_z^0 + E_z^2 = 0 \quad \text{on} \quad \mathcal{M},$$

$$E_z^1 = E_z^2, \quad \frac{\partial E_z^1}{\partial r} = \frac{\partial E_z^2}{\partial r} \quad \text{on} \quad \mathcal{N},$$

and, consequently, $b_n H_n^{(1)}(kR) = c_n J_n(kR) = d_n$, then for new unknowns $d_n$ we have a paired series functional equation

$$\sum_{n=-\infty}^{+\infty} d_n e^{in\alpha} = -E_z^0(R, \alpha) \quad \text{on} \quad \mathcal{M},$$

$$\sum_{n=-\infty}^{+\infty} d_n \left[ \frac{H_n^{(1)'}(kR)}{H_n^{(1)}(kR)} - \frac{J_n'(kR)}{J_n(kR)} \right] e^{in\alpha} = 0 \quad \text{on} \quad \mathcal{N}.$$

The work [16] specifies two ways to reduce PSFE to ISLAE

$$-2\pi d_k + \frac{1}{2\pi} \sum_{n=-\infty}^{+\infty} d_n \gamma_n \sum_{m=-\infty}^{+\infty} \frac{1}{\gamma_m} I_{n-m} J_{m-k} = \sum_{n=-\infty}^{+\infty} e_n^0 I_{n-k}, \quad k = 0, \pm 1, \ldots, \quad (32)$$

It is easy to calculate integrals

$$I_n = \int_M e^{in\beta} \, d\beta, \quad J_n = \int_N e^{in\beta} \, d\beta.$$

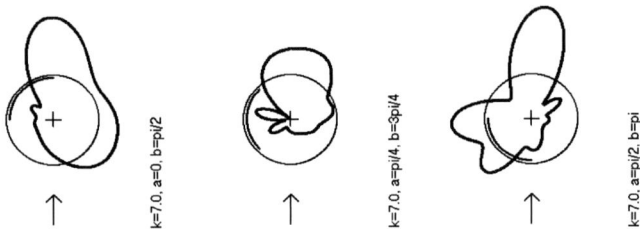

Figure 6. Scattering diagram for different positions of the strip.

Figure 6 shows the scattering diagrams of the field in the far zone for different positions of the strip.

## 6. Electromagnetic Wave Scattering by the Axis-Symmetric Body

If we move on from Maxwell equations in spherical coordinates $(r, \theta, \alpha)$ to Debye potentials or to Hertz vectors, then particular solutions of these equations can be found by the separating variables method [2], [18].

### 6.1. Wave Scattering by a Dielectric Body

Let us limit ourselves to the case when the electromagnetic field does not depend on the coordinate $\alpha$. Then the components of the waves outgoing to infinity have the form

$$E_r(r, \theta) = \sum_{n=1}^{+\infty} \frac{n(n+1)}{r} d_n \zeta_n^{(2)}(kr) \varphi_n(\theta),$$

$$E_\theta(r,\theta) = \sum_{n=1}^{+\infty} d_n \frac{1}{r} \frac{d}{dr}\left(r\zeta_n^{(2)}(kr)\right) \varphi_n'(\theta),$$

$$E_\varphi(r,\theta) = i\omega\mu_0\mu \sum_{n=1}^{+\infty} b_n \zeta_n^{(2)}(kr) \varphi_n'(\theta),$$

$$H_r(r,\theta) = \sum_{n=1}^{+\infty} \frac{n(n+1)}{r} b_n \zeta_n^{(2)}(kr) \varphi_n(\theta),$$

$$H_\theta(r,\theta) = \sum_{n=1}^{+\infty} b_n \frac{1}{r} \frac{d}{dr}\left(r\zeta_n^{(2)}(kr)\right) \varphi_n'(\theta),$$

$$H_\varphi(r,\theta) = -i\omega\varepsilon_0\varepsilon \sum_{n=1}^{+\infty} d_n \zeta_n^{(2)}(kr) \varphi_n'(\theta),$$

where $\zeta_n^{(2)}(z)$ are Hankel spherical functions and $\varphi_n(\theta)$ are Legendre joined functions,

$$\zeta_n^{(2)}(z) = \sqrt{\frac{\pi}{2z}} H_{n+1/2}^{(2)}(z), \quad \varphi_n(\theta) = P_n^{(m)}(\cos\theta).$$

In a bounded region instead of spherical Hankel functions we should take the spherical Bessel functions $\psi_n(z)$ (it is Sommerfeld notation). The coefficients of the series in this case will be denoted by $a_n$ and $c_n$ (instead of $b_n$ and $d_n$).

In the work [17] it is shown that on the boundary of the dielectric axis-symmetric body, the surface of which is given by the function $r = g(\theta)$, the expressions $g'(\theta) E_r + g(\theta) E_\theta$, $E_\varphi$, $g'(\theta) H_r + g(\theta) H_\theta$ $H_\varphi$ should be continuous. If the components of the wave running on the body are also independent of the coordinate $\alpha$, then continuity conditions of expressions $g'(\theta) H_r + g(\theta) H_\theta$ and $E_\varphi$ for $r = g(\theta)$ give series equations for determining the coefficients $a_n$, $b_n$, and the continuity conditions of expressions $g'(\theta) E_r + g(\theta) E_\theta$ and $H_\varphi$ give series equations for determining the coefficients $c_n$, $d_n$. The first pair of equations has the form

$$\sum_{n=1}^{+\infty} A_n^I(\theta) a_n + \sum_{n=1}^{+\infty} B_n^I(\theta) b_n = C^I(\theta), \quad 0 < \theta < \pi, \quad (33)$$

$$\sum_{n=1}^{+\infty} A_n^{II}(\theta) a_n + \sum_{n=1}^{+\infty} B_n^{II}(\theta) b_n = C^{II}(\theta), \quad 0 < \theta < \pi, \quad (34)$$

where

$$A_n^I(\theta) = \frac{g'(\theta)}{g(\theta)} n(n+1) \psi_n(k_2 g(\theta)) \varphi_n(\theta) + \left(\psi_n(k_2 g(\theta)) + k_2 g(\theta) \psi_n'(k_2 g(\theta))\right) \varphi_n'(\theta)$$

$$B_n^I(\theta) = -\frac{g'(\theta)}{g(\theta)} n(n+1) \zeta_n^{(2)}(k_1 g(\theta)) \psi_n(k_2 r)$$
$$- \left(\zeta_n^{(2)}(k_1 g(\theta)) + k_1 g(\theta) \zeta_n^{(2)\prime}(k_1 g(\theta))\right) \psi_n'(k_2 r),$$

$$A_n^{II}(\theta) = i\omega\mu_0\mu_2 \psi_n(k_2 g(\theta)) \psi_n'(k_2 r), \quad B_n^{II}(\theta) = -i\omega\mu_0\mu_1 \zeta_n^{(2)}(k_1 g(\theta)) \psi_n'(k_2 r),$$
$$C^I(\theta) = g'(\theta) H_r^0(g(\theta), \theta) + g(\theta) H_\theta^0(g(\theta), \theta), \quad C^{II}(\theta) = E_\varphi^0(g(\theta), \theta).$$

If we multiply the series equations by $P_m(\cos\theta) \sin\theta$ and integrate them by $\theta$ from 0 to $\pi$, we get an ISLAE of the form

$$\sum_{n=1}^{+\infty} A_{nm}^I a_n + \sum_{n=1}^{+\infty} B_{nm}^I b_n = C_m^I, \quad m = 1, 2, \ldots, \tag{35}$$

$$\sum_{n=1}^{+\infty} A_{nm}^{II} a_n + \sum_{n=1}^{+\infty} B_{nm}^{II} b_n = C_m^{II}, \quad m = 1, 2, \ldots \tag{36}$$

Here $A_{nm}^I$, $B_{nm}^I$, $A_{nm}^{II}$, $B_{nm}^{II}$, $C_m^I$, $C_m^{II}$ are integrals of the functions $A_n^I(\theta)$, $B_n^I(\theta)$, $A_n^{II}(\theta)$, $B_n^{II}(\theta)$, $C^I(\theta)$, $C^{II}(\theta)$, multiplied by $P_m(\cos\theta) \sin\theta$. Similar equations are derived for unknowns $c_n$ and $d_n$.

## 6.2. Wave Scattering by a Thin Conductive Spherical Ring

Finally, we consider the problem of diffraction of a spherical wave by a thin conductive spherical ring, which is a part of the sphere of the radius $R$ for values of latitude $\theta$ from $\alpha$ to $\beta$, using some results of work [17].

In this case, the field component has the form:

$$E_\varphi^1 = i\omega\mu_0\mu \sum_{n=1}^{+\infty} b_n \zeta_n^{(2)}(kr) \varphi_n'(\theta), \quad H_\theta^1 = \sum_{n=1}^{+\infty} b_n \frac{1}{r} \left(\zeta_n^{(2)}(kr) + kr \zeta_n^{(2)\prime}(kr)\right) \varphi_n'(\theta)$$

out of the sphere and

$$E_\varphi^2 = i\omega\mu_0\mu \sum_{n=1}^{+\infty} a_n \psi_n(kr) \varphi_n'(\theta), \quad H_\theta^2 = \sum_{n=1}^{+\infty} a_n \frac{1}{r} \left(\psi_n(kr) + kr \psi_n'(kr)\right) \varphi_n'(\theta)$$

inside sphere. The boundary conditions $E^1_\varphi + E^0_\varphi = 0$, $E^2_\varphi + E^0_\varphi = 0$ should be fulfilled on $\mathcal{M} = [\alpha, \beta]$ when $r = R$. The conjugation conditions $E^1_\varphi = E^2_\varphi$, $H^1_\theta = H^2_\theta$ should be fulfilled on the supplement $\mathcal{N}$ of the segment $\mathcal{M}$ up to the segment $[0, \pi]$.

For the new unknown coefficients

$$c_n = b_n\, \zeta^{(2)}_n(kR) = a_n\, \psi_n(kR).$$

we have PSFE

$$i\omega\mu_0\mu \sum_{n=1}^{+\infty} c_n \varphi'_n(\theta) = -E^0_\varphi \quad \text{on} \quad \mathcal{M},$$

$$\sum_{n=1}^{+\infty} c_n \gamma_n \varphi'_n(\theta) = 0 \quad \text{on} \quad \mathcal{N}, \quad \gamma_n = \frac{\zeta^{(2)\prime}_n(kR)}{\zeta^{(2)}_n(kR)} - \frac{\psi'_n(kR)}{\psi_n(kR)}$$

(if the value $kR$ is not a zero for one of the functions $\psi_n(\cdot)$).

We suppose that a spherical wave run over the ring from the source at the point $(0, 0, z_0)$. If we multiply the series equations by $\varphi'_m(\theta) \sin\theta$ and integrate them by $\theta$ on sets $\mathcal{M}$ and $\mathcal{N}$ respectively, then we get ISLAE ($b^0_l = 1$)

$$\sum_{n=1}^{+\infty} c_n I_{nm} = -F_m, \quad \sum_{n=1}^{+\infty} c_n \gamma_n J_{nm} = 0, \qquad (37)$$

where

$$I_{nm} = \int_{\mathcal{M}} \varphi'_n(\theta)\, \varphi'_m(\theta) \sin\theta d\theta, \quad J_{nm} = p_n \delta_{nm} - I_{nm}, \quad p_n = \frac{(n+1)n}{n+1/2},$$

$$F_m = \int_{\mathcal{M}} \zeta^{(2)}(k\tilde{r})\, \varphi'_l(\tilde{\theta})\, \varphi'_m(\theta) \sin\theta\, d\theta.$$

The expressions $\tilde{r}$ and $\tilde{\theta}$ are local spherical coordinates,

$$\tilde{r} = \sqrt{r^2 - 2rz_0\cos\theta + z_0^2}, \quad \cos\tilde{\theta} = \frac{r\cos\theta - z_0}{\tilde{r}}, \quad \sin\tilde{\theta} = \frac{r\sin\theta}{\tilde{r}}, \quad \tilde{\varphi} = \varphi.$$

If we rewrite the ISLAE in the form

$$p_k c_k - \sum_{m=1}^{+\infty} c_m J_{mk} = -F_k, \quad p_m c_m \gamma_m = \sum_{n=1}^{+\infty} c_n \gamma_n I_{nm} = 0,$$

then finally we get

$$p_k c_k - \sum_{n=1}^{+\infty} c_n \gamma_n \sum_{m=1}^{+\infty} \frac{1}{p_m \gamma_m} I_{nm} J_{mk} = -F_k, \quad k = 1, 2, \ldots \quad (38)$$

If a spherical wave runs over a conductive sphere, then $\mathcal{M} = [0, \pi]$, $\mathcal{N} = \emptyset$. In this case, the ISLAE degenerates into a set of simple equations $p_k \zeta_k^{(2)}(kR) b_k = -F_k, \ k = 1, 2, \ldots$.

## 6.3. Computational Experiment

Here are some results of computational experiments.

Figure 7 shows scattering diagrams for bodies with a periodical non-coordinate boundary $g(\theta) = R + 0.2 \sin m\theta$ when $m = 2$, $m = 3$ and $m = 4$. The value $R$ is equal to the wavelength.

The scattering diagrams for the different positions of thin conductive ring for cases $[0, \pi/3]$, $[\pi/3, 2\pi/3]$, $[2\pi/3, \pi]$ are shown on the Figure 8.

We denote that spherical ring of the large size is a spherical resonator with one or two holes. For some frequencies of the external electromagnetic field (close to the resonator's eigen frequencies) it should be expected that the characteristics of the field inside the ring can change sharply. A computational experiment confirms this effect.

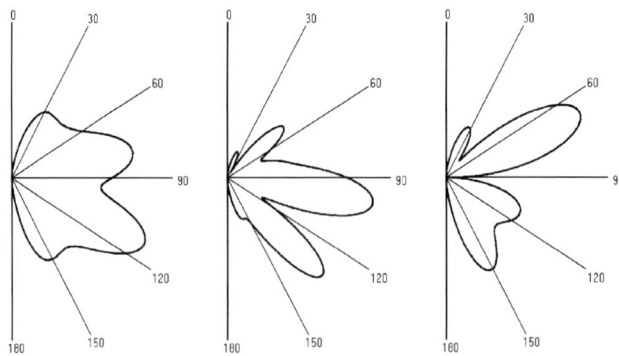

Figure 7. The intensity of the field of bodies with a non-coordinate boundary.

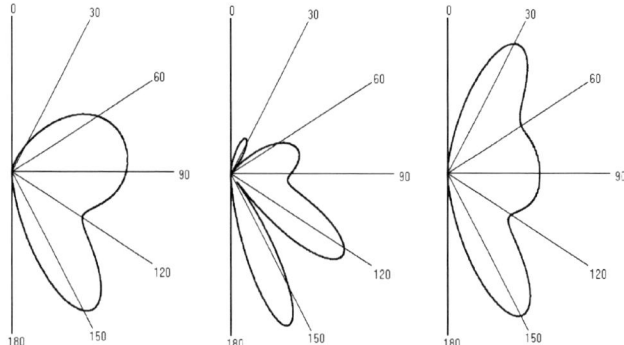

Figure 8. The intensity of the scattering field of spherical ring.

## Acknowledgments

This paper has been supported by the Kazan Federal University Strategic Academic Leadership Program (PRIORITY-2030).

## References

[1] Jackson, J. D. *Classical Electrodynamics*. New York: John Wiley & Sons, Inc., 1998.

[2] Hönl, H., Maue, A. W., and Westpfahl, K. *Theorie der Beugung*. Berlin: Springer-Verlag, 1961. [in German]

[3] Colton D., and Kress R. *Integral Equation Methods in Scattering Theory*. New York: John Wiley & Sons, Inc., 1983.

[4] Galishnikova T. N., and Il'inskii, A. S. *The Integral Equation Method in Problems of Wave Diffraction*. Moscow: MAKS Press, 2013. [in Russian].

[5] Hsiao, G. C., and Wendland, W. L. *Boundary Integral Equations*. New York: Springer, 2008.

[6] Costabel, M., Darrigrand, E., and Kone, E.H. Volume and Surface Integral Equations for Electromagnetic Scattering by a Dielectric Body.

*Journal of Computational and Applied Mathematics.* 2010. Vol. 234, No. 6. 1817-1825.

[7] Costabel, M., and Le Louer, F. On the Kleinman-Martin Integral Equation Method for Electromagnetic Scattering by a Dielectric Body. *SIAM Journal on Applied Mathematics.* 2011. Vol. 71, No 2. 635-656.

[8] Duran, M., Muga, I., and Nédélec, J.-C. The Helmholtz Equation in a Locally Perturbed Half-Space with Non-Absorbing Boundary. *Archive for Rational Mechanics and Analysis.* 2009. Vol. 191, No. 1. 143-172.

[9] Ilyinsky, A. S., and Smirnov, Yu. G. *Electromagnetic Wave Diffraction by Conductiong Screens.* Utrecht: VSP, 1998.

[10] Smirnov, Y. G., and Tsupak, A. A. Existence and Uniqueness Theorems in Electromagnetic Diffraction on Systems of Lossless Dielectrics and Perfectly Conducting Screens. *Applicable Analysis.* 2017. Vol. 96, No. 8. 1326-1341.

[11] Samokhin, A. B. *Integral Equations and Iteration Methods in Electromagnetic Scattering.* Utrecht: VSP, 2001.

[12] Samokhin, A. B. *Volume Singular Integral Equations of Electrodynamics.* Moscow: Tekhnosphera, 2021. [in Russian]

[13] Mittra, R., and Lee, S. W. *Analytical Techniques in the Theory of Guided Waves.* New York: The Macmillan Company, 1971.

[14] Pleshchinskii, I., and Pleshchinskii, N. Parallel Algorithm of Solving the Problem of the Electromagnetic Wave Diffraction by the Curvilinear Media Interface. *Journal of Advanced Research in Dynamical and Control Systems.* 2018. Vol. 10, Is. 10 Special Issue. 1741-1746.

[15] Pleshchinskaya, I. E, and Pleshchinskii, N. B. Infinite Sets of Linear Algebraic Equations in the Problems of Diffraction of Electromagnetic Waves by the Non-Coordinate Periodic Media Interfaces. *Lobachevskii Journal of Mathematics.* 2019. Vol. 40, Is. 10. 1685-1694.

[16] Pleshchinskaya, I. E., and Pleshchinskii, N. B. On the Scattering of Electromagnetic Waves by Cylindrical Bodies with Non-Coordinate Boundaries. *Lobachevskii Journal of Mathematics*. 2020. Vol. 41, No. 7. 1385-1395.

[17] Pleshchinskaya, I. E., and Pleshchinskii, N. B. On the Scattering of Electromagnetic Waves by Axial-Symmetric Bodies. *Lobachevskii Journal of Mathematics*. 2021. Vol. 42, No. 6. 1381-1390.

[18] Angot, A. *Compléments de Mathématiques a l'Usage des Ingénieurs de l'Électrotechnique et des Télécommunications*. Paris, 1957. [in French]

[19] Pleshchinskii, N. B. On Generalized Solutions of Problems of Electromagnetic Wave Diffraction by Screens in the Closed Cylindrical Waveguides. *Lobachevskii Journal of Mathematics*. 2019. Vol.40, Is. 2. 201-209.

[20] Pleshchinskii, N. B. On Generalized Solutions of the Problems of Electromagnetic Wave Diffraction in the Open Space. *Lobachevskii Journal of Mathematics*. 2021. Vol. 42, No. 6. 1391-1401.

# Bibliography

**Advanced electromagnetic wave propagation methods**
| | |
|---|---|
| *LCCN* | 2021025798 |
| *Type of material* | Book |
| *Personal name* | Gonzalez, Guillermo, 1944- author. |
| *Main title* | Advanced electromagnetic wave propagation methods / Guillermo Gonzalez. |
| *Edition* | First edition. |
| *Published/Produced* | Boca Raton: CRC Press, [2022] |
| *Description* | 1 online resource |
| *ISBN* | 9781000476675 (epub) |
| | 9781003219729 (ebk) |
| | (hbk) |
| | (pbk) |
| *LC classification* | QC665.T7 |
| *Summary* | "This textbook provides a solid foundation into the approaches used in the analysis of complex electromagnetic problems and wave propagation. The techniques discussed are essential to obtain closed-form solutions or asymptotic solutions and meet an existing need for instructors and students in electromagnetic theory"-- Provided by publisher. |
| *LC Subjects* | Electromagnetic waves--Transmission--Mathematical models. |
| | Electromagnetic theory Mathematics. |
| | Radio wave propagation. |
| *Notes* | Includes bibliographical references and index. |
| *Additional formats* | Print version: Gonzalez, Guillermo, 1944- Advanced electromagnetic wave propagation methods First edition. Boca Raton: CRC Press, [2022] 9781032113708 (DLC) 2021025797 |

**Advanced electromagnetic wave propagation methods**
| | |
|---|---|
| *LCCN* | 2021025797 |
| *Type of material* | Book |
| *Personal name* | Gonzalez, Guillermo, 1944- author. |

|   |   |
|---|---|
| *Main title* | Advanced electromagnetic wave propagation methods / Guillermo Gonzalez. |
| *Edition* | First edition. |
| *Published/Produced* | Boca Raton: CRC Press, [2022] |
| *ISBN* | 9781032113708 (hbk) |
|  | 9781032114002 (pbk) |
|  | (ebk) |
| *LC classification* | QC665.T7 G67 2022 |
| *Summary* | "This textbook provides a solid foundation into the approaches used in the analysis of complex electromagnetic problems and wave propagation. The techniques discussed are essential to obtain closed-form solutions or asymptotic solutions and meet an existing need for instructors and students in electromagnetic theory"-- Provided by publisher. |
| *LC Subjects* | Electromagnetic waves--Transmission--Mathematical models. |
|  | Electromagnetic theory Mathematics. |
|  | Radio wave propagation. |
| *Notes* | Includes bibliographical references and index. |
| *Additional formats* | Online version: Gonzalez, Guillermo, 1944- Advanced electromagnetic wave propagation methods Boca Raton: CRC Press, 2022 9781003219729 (DLC) 2021025798 |

## Advances in millimeter wave technology

|   |   |
|---|---|
| *LCCN* | 2019347576 |
| *Type of material* | Book |
| *Personal name* | Shukla, Awadesh Kumar, author. |
| *Main title* | Advances in millimeter wave technology / Awadesh Kumar Shukla. |
| *Published/Produced* | New Delhi: Defence Research and Development Organisation, Ministry of Defence, 2019. |
| *Description* | viii, 208 pages: illustrations (black and white, and color); 24 cm. |
| *ISBN* | 9788186514634 |
| *LC classification* | TK7876.5 .S54 2019 |
| *Related names* | Defence Research & Development Organisation (India), copyright holder, publisher. |

| | |
|---|---|
| *LC Subjects* | Millimeter waves.<br>Electromagnetic waves.<br>Microwaves--Military applications.<br>Electronics in military engineering. |
| *Notes* | Includes bibliographical references and index. |
| *Reproduction no./Source* | Library of Congress - New Delhi Overseas Office |
| *Series* | DRDO monographs/special publications series |

## Analysis and modeling of radio wave propagation

| | |
|---|---|
| *LCCN* | 2016045806 |
| *Type of material* | Book |
| *Personal name* | Coleman, Christopher, 1950- author. |
| *Main title* | Analysis and modeling of radio wave propagation / Christopher John Coleman, University of Adelaide. |
| *Published/Produced* | Cambridge, United Kingdom; New York, NY: Cambridge University Press, [2017]<br>©2016 |
| *Links* | Contributor biographical information https://www.loc.gov/catdir/enhancements/fy1701/2016045806-b.html<br>Publisher description https://www.loc.gov/catdir/enhancements/fy1701/2016045806-d.html<br>Table of contents only https://www.loc.gov/catdir/enhancements/fy1701/2016045806-t.html |
| *ISBN* | 9781107175563 (hardback; alk. paper)<br>1107175569 (hardback; alk. paper) |
| *LC classification* | TK6553 .C635 2017 |
| *Summary* | "This comprehensive guide helps readers understand the theory and techniques needed to analyze and model radio wave propagation in complex environments. All of the essential topics are covered, from the fundamental concepts of radio systems, to complex propagation phenomena. These topics include diffraction, ray tracing, scattering, atmospheric ducting, ionospheric ducting, scintillation, and propagation through both urban and non-urban environments. Emphasis is placed on practical procedures, with detailed discussion of |

|  |  |
|---|---|
|  | numerical and mathematical methods providing readers with the necessary skills to build their own propagation models and develop their own techniques. MATLAB functions illustrating key modeling ideas are provided online. This is an invaluable resource for anyone wanting to use propagation models to understand the performance of radio systems for navigation, radar, communications, or broadcasting"-- Provided by publisher. |
| *Contents* | Basic concepts page - The fundamentals of electromagnetic waves - The reciprocity, compensation and extinction theorems - The effect of obstructions upon radio wave propagation - Geometric optics - Propagation through irregular media - The approximate solution of Maxwell's equations - Propagation in the ionospheric duct - Propagation in the lower atmosphere - Transionospheric propagation and scintillation. |
| *LC Subjects* | Radio wave propagation. Radio wave propagation--Mathematical models. Electromagnetic waves. |
| *Notes* | Includes bibliographical references and index. |

| **Anechoic and reverberation chambers: theory, design and measurements** | |
|---|---|
| *LCCN* | 2018050886 |
| *Type of material* | Book |
| *Personal name* | Xu, Qian, 1985- author. |
| *Main title* | Anechoic and reverberation chambers: theory, design and measurements / Qian Xu, College of Electronic and Information Engineering, Nanjing University of Aeronautics and Astronautics, Nanjing, China, Yi Huang, The University of Liverpool, Liverpool, UK. |
| *Published/Produced* | Hoboken, NJ, USA: Wiley-IEEE Press, 2018. |
| *Description* | 1 online resource. |
| *ISBN* | 9781119362029 (Adobe PDF) 9781119362043 (ePub) |
| *LC classification* | QC667.A54 |
| *Related names* | Huang, Yi, 1964- author. |

| | |
|---|---|
| *LC Subjects* | Anechoic chambers. |
| | Electromagnetic reverberation chambers. |
| | Electromagnetic measurements. |
| | Electromagnetic waves--Transmission. |
| | Shielding (Electricity) |
| *Notes* | Includes bibliographical references and index. |
| *Additional formats* | Print version: Xu, Qian, 1985- author. Anechoic and reverberation chambers Hoboken, NJ, USA: Wiley-IEEE Press, 2018 9781119361688 (DLC) 2018025829 |

## Anechoic and reverberation chambers: theory, design and measurements

| | |
|---|---|
| *LCCN* | 2018025829 |
| *Type of material* | Book |
| *Personal name* | Xu, Qian, 1985- author. |
| *Main title* | Anechoic and reverberation chambers: theory, design and measurements / Qian Xu, College of Electronic and Information Engineering, Nanjing University of Aeronautics and Astronautics, Nanjing, China, Yi Huang, The University of Liverpool, Liverpool, UK. |
| *Published/Produced* | Hoboken, NJ, USA: Wiley-IEEE Press, 2018. |
| *ISBN* | 9781119361688 (hardcover) |
| *LC classification* | QC667.A54 X83 2018 |
| *Related names* | Huang, Yi, 1964- author. |
| *LC Subjects* | Anechoic chambers. |
| | Electromagnetic reverberation chambers. |
| | Electromagnetic measurements. |
| | Electromagnetic waves--Transmission. |
| | Shielding (Electricity) |
| *Notes* | Includes bibliographical references and index. |
| *Additional formats* | Online version: Xu, Qian, 1985- author. Anechoic and reverberation chambers Hoboken, NJ, USA: Wiley-IEEE Press, 2018 9781119362029 (DLC) 2018050886 |

## Antennas: rigorous methods of analysis and synthesis

| | |
|---|---|
| *LCCN* | 2020041783 |
| *Type of material* | Book |

| | |
|---|---|
| *Personal name* | Levin, Boris, 1937- author. |
| *Main title* | Antennas: rigorous methods of analysis and synthesis / Boris Levin, Holon Institute of Technology, Israel. |
| *Published/Produced* | Boca Raton: CRC Press, Taylor & Francis Group, [2021] |
| *Description* | vii, 314 pages; 25 cm |
| *ISBN* | 9780367489236 (hardcover) |
| *LC classification* | TK6565.A6 L455 2021 |
| *Summary* | "The book contains new variant of solving the equation of Leontovich, which is most rigorous equation for the current in a thin linear antenna. This solution based in particular on the method of constants variation permits to calculate the sum of any members of series, as well as the total sum of the series members, i.e., solve a problem that has agitated specialists for more than half a century. The book gives a consistent exposition of the theory of integral equations and discusses the features of their application for analyzing different types of antennas. The book considers new ways of analyzing antennas: method of calculating an antenna gain based on its main radiation patterns, procedure of calculating the directional characteristics of radiators with known distribution of current amplitude, method of electrostatic analogy which permits to compare the electromagnetic fields of high-frequency currents with electrostatic charges fields located on linear conductors and to improve the directional characteristics of log-periodic and director-type antennas"-- Provided by publisher. |
| *LC Subjects* | Antennas (Electronics)--Mathematical models. Antenna radiation patterns--Mathematical models. Electromagnetic waves--Mathematical models. |
| *Notes* | "A Science Publishers book." Includes bibliographical references (pages 308-311) and index. |

**Antennas: rigorous methods of analysis and synthesis**
*LCCN*               2020756514

# Bibliography

| | |
|---|---|
| *Type of material* | Book |
| *Personal name* | Levin, Boris, 1937- author. |
| *Main title* | Antennas: rigorous methods of analysis and synthesis / Boris Levin, Holon Institute of Technology, Israel. |
| *Published/Produced* | Boca Raton: CRC Press, Taylor and Francis Group, [2021] |
| *Description* | 1 online resource (vii, 314 pages) |
| *ISBN* | 9781000220056 ebook (hardcover) |
| *LC classification* | TK6565.A6 |
| *Summary* | "The book contains new variant of solving the equation of Leontovich, which is most rigorous equation for the current in a thin linear antenna. This solution based in particular on the method of constants variation permits to calculate the sum of any members of series, as well as the total sum of the series members, i.e., solve a problem that has agitated specialists for more than half a century. The book gives a consistent exposition of the theory of integral equations and discusses the features of their application for analyzing different types of antennas. The book considers new ways of analyzing antennas: method of calculating an antenna gain based on its main radiation patterns, procedure of calculating the directional characteristics of radiators with known distribution of current amplitude, method of electrostatic analogy which permits to compare the electromagnetic fields of high-frequency currents with electrostatic charges fields located on linear conductors and to improve the directional characteristics of log-periodic and director-type antennas"-- Provided by publisher. |
| *LC Subjects* | Antennas (Electronics)--Mathematical models. Antenna radiation patterns--Mathematical models. Electromagnetic waves--Mathematical models. |
| *Notes* | "A Science Publishers book." Includes bibliographical references (pages 308-311) and index. |

| | |
|---|---|
| *Additional formats* | Print version: Antennas Boca Raton: CRC Press, Taylor & Francis Group, [2021] 9780367489236 (DLC) 2020041783 |

*Computational methods for electromagnetic inverse scattering*

| | |
|---|---|
| *LCCN* | 2017045207 |
| *Type of material* | Book |
| *Personal name* | Chen, Xudong, 1976- author. |
| *Main title* | Computational methods for electromagnetic inverse scattering / by Xudong Chen. |
| *Published/Produced* | Singapore; Hoboken, NJ: John Wiley & Sons, 2018. |
| *ISBN* | 9781119311980 (cloth) |
| *LC classification* | QC665.S3 C473 2018 |
| *Contents* | Fundamentals of electromagnetic wave theory - Time reversal imaging - Inverse scattering problems of small scatterers - Linear sampling method - Reconstructing dielectric scatterers - Reconstructing perfect electric conductors - Inversion for phaseless data - Inversion with an inhomogeneous background medium - Resolution of computational imaging. |
| *LC Subjects* | Electromagnetic waves--Scattering--Mathematical models. |
| | Scattering (Physics)--Mathematics. |
| | Inverse scattering transform. |
| *Notes* | Includes index. |
| *Additional formats* | Online version: Chen, Xudong, 1976- author. Computational methods for electromagnetic inverse scattering Singapore; Hoboken, NJ: John Wiley & Sons, 2018 9781119312017 (DLC) 2017058095 |

**Computational methods for electromagnetic inverse scattering**

| | |
|---|---|
| *LCCN* | 2017058095 |
| *Type of material* | Book |
| *Personal name* | Chen, Xudong, 1976- author. |
| *Main title* | Computational methods for electromagnetic inverse scattering / by Xudong Chen. |
| *Published/Produced* | Singapore; Hoboken, NJ: John Wiley & Sons, 2018. |
| *Description* | 1 online resource. |
| *ISBN* | 9781119312017 (pdf) |

|   |   |
|---|---|
| | 9781119312000 (epub) |
| *LC classification* | QC665.S3 |
| *Contents* | Fundamentals of electromagnetic wave theory - Time reversal imaging - Inverse scattering problems of small scatterers - Linear sampling method - Reconstructing dielectric scatterers - Reconstructing perfect electric conductors - Inversion for phaseless data - Inversion with an inhomogeneous background medium - Resolution of computational imaging. |
| *LC Subjects* | Electromagnetic waves--Scattering--Mathematical models. |
| | Scattering (Physics)--Mathematics. |
| | Inverse scattering transform. |
| *Notes* | Includes index. |
| *Additional formats* | Print version: Chen, Xudong, 1976- author. Computational methods for electromagnetic inverse scattering Singapore; Hoboken, NJ: John Wiley & Sons, 2018 9781119311980 (DLC) 2017045207 |

**Computing the flow of light: nonstandard FDTD methodologies for photonics design**

|   |   |
|---|---|
| *LCCN* | 2016033109 |
| *Type of material* | Book |
| *Personal name* | Cole, James B. (James Bradford), author. |
| *Main title* | Computing the flow of light: nonstandard FDTD methodologies for photonics design / James B. Cole, Saswatee Banerjee. |
| *Published/Produced* | Bellingham, Washington, USA: SPIE Press, [2017] ©2017 |
| *Description* | xvi, 413 pages: illustrations (some color); 26 cm + 1 CD-ROM (4 3/4 in.) |
| *ISBN* | 9781510604810 (softcover) |
| | 1510604812 (softcover) |
| | (pdf) |
| | (pdf) |
| | (epub) |
| | (epub) |
| | (Kindle/mobi) |
| | (Kindle/mobi) |

| | |
|---|---|
| *LC classification* | TA1522 .C65 2017 |
| *Related names* | Banerjee, Saswatee, author. |
| *Summary* | "Finite difference time domain (FDTD) computes the time evolution of a system at discrete time steps, and the resulting periodic visualization yields insight into the system. FDTD and FDTD-like methods can be used to solve a wide variety of problems, including-- but not limited to--the wave equation, Maxwell's equations, and the Schrödinger equation. In addition to introducing useful new methodologies, this book provides readers with analytical background and simulation examples that will help them develop their own methodologies to solve yet-to-be-posed problems. The book is written for students, engineers, and researchers grappling with problems that cannot be solved analytically. It could also be used as a textbook for a mathematical physics or engineering class"-- Provided by publisher. |
| *Contents* | Finite difference approximations - Accuracy, stability and convergence of numerical algorithms - Introduction - Finite difference models of the simple harmonic oscillator - The one-dimensional wave equation - Finite difference time domain algorithms for the one-dimensional wave equation - Program development and applications of finite difference time domain algorithms in one-dimension - Finite difference time domain algorithms to solve the wave equation in two and three dimensions - Review of electromagnetic theory - The Yee algorithm in one dimension - The Yee algorithm in two and three dimensions - Example applications of FDTD - FDTD for dispersive materials - Photonics problems - Photonics design. |
| *LC Subjects* | Photonics--Mathematics. |
| | Electromagnetic waves--Mathematical models. |
| | Finite differences. |
| | Time-domain analysis. |
| *Notes* | Includes bibliographical references and index. |

**Department of Defense electromagnetic spectrum operations: challenges and opportunities in the invisible battlespace: hearing before the Subcommittee on Cyber, Innovative Technologies, and Information Systems of the Committee on Armed Services, House of Representatives, One Hundred Seventeenth Congress, first session, hearing held March 19, 2021.**

| | |
|---|---|
| *LCCN* | 2020440196 |
| *Type of material* | Book |
| *Corporate name* | United States. Congress. House. Committee on Armed Services. Subcommittee on Cyber, Innovative Technologies, and Information Systems, author. |
| *Main title* | Department of Defense electromagnetic spectrum operations: challenges and opportunities in the invisible battlespace: hearing before the Subcommittee on Cyber, Innovative Technologies, and Information Systems of the Committee on Armed Services, House of Representatives, One Hundred Seventeenth Congress, first session, hearing held March 19, 2021. |
| *Published/Produced* | Washington: U.S. Government Publishing Office, 2021. |
| *Description* | iii, 88 pages: illustrations; 24 cm |
| *LC classification* | KF27 .A73597 2021d |
| *Portion of title* | Challenges and opportunities in the invisible battlespace |
| *LC Subjects* | Defense industries--Technological innovations--United States. Electromagnetic waves. Weapons systems--Technological innovations--United States. Electronics in military engineering. National security--United States. |
| *Other Subjects* | Defense industries--Technological innovations. Electromagnetic waves. Electronics in military engineering. National security. Weapons systems--Technological innovations. United States. |
| *Form/Genre* | Legislative hearings. |

|              |              |
|---|---|
| *Notes* | Legislative hearings.<br>Shipping list no.: 2022-0037-P.<br>H.A.S.C. no. 117-12.<br>Date of hearing: 2021-03-19. |
| *Additional formats* | Online version: United States. Congress. House. Committee on Armed Services. Subcommittee on Cyber, Innovative Technologies, and Information Systems. Department of Defense electromagnetic spectrum operations (OCoLC)1280153763 |

**Elastic scattering of electromagnetic radiation: analytic solutions in diverse backgrounds**

|              |              |
|---|---|
| *LCCN* | 2017040817 |
| *Type of material* | Book |
| *Personal name* | Sharma, Subodh K. (Subodh Kumar K.), author. |
| *Main title* | Elastic scattering of electromagnetic radiation: analytic solutions in diverse backgrounds / Subodh Kumar Sharma. |
| *Published/Produced* | Boca Raton, FL: CRC Press, Taylor & Francis Group, [2018]<br>©2018 |
| *ISBN* | 9781498748575 (hardback; alk. paper)<br>1498748570 (hardback; alk. paper)<br>(ebook)<br>(ebook) |
| *LC classification* | QC665.S3 S53 2018 |
| *Summary* | "The book examines analytic solutions of the scattering problems of types: Scattering by single particle and tenuous system of uncorrelated scatterers in diverse backgrounds"-- Provided by publisher. |
| *Contents* | Single particle scattering - Approximate formulas - Scattering by an assembly of particles. |
| *LC Subjects* | Electromagnetic waves--Scattering--Mathematics.<br>Elastic scattering. |
| *Notes* | Includes bibliographical references and index. |

**Electromagnetic and acoustic wave tomography: direct and inverse problems in practical applications**

|              |              |
|---|---|
| *LCCN* | 2018006943 |

# Bibliography

| | |
|---|---|
| *Type of material* | Book |
| *Main title* | Electromagnetic and acoustic wave tomography: direct and inverse problems in practical applications / edited by Nathan Blaunstein, Vladimir Yakubov. |
| *Published/Produced* | Boca Raton: CRC Press, Taylor & Francis Group, [2019] |
| *Description* | xix, 360 pages; 25 cm |
| *ISBN* | 9781138490734 (hb: acid-free paper) ebook |
| *LC classification* | TA1560 .E44 2019 |
| *Related names* | Blaunstein, Nathan, editor. Yakubov, Vladimir, editor. |
| *Summary* | This book discusses the development of radio-wave tomography methods as a means of remote non-destructive testing, diagnostics of the internal structure of semi-transparent media, and reconstruction of the shapes of opaque objects based on multi-angle sounding. It describes physical-mathematical models of systems designed to reconstruct images of hidden objects, based on tomographic processing of multi-angle remote measurements of scattered radio and acoustic (ultrasonic) wave radiation-- Provided by publisher. |
| *LC Subjects* | Three-dimensional imaging--Mathematics. Tomography. Acoustic emission testing. Electromagnetic waves--Mathematical models. Remote sensing. Radar. |
| *Notes* | "A CRC title, part of the Taylor & Francis imprint, a member of the Taylor & Francis Group, the academic division of T&F Informa plc." Includes bibliographical references and index. |

**Electromagnetic and acoustic wave tomography: direct and inverse problems in practical applications**

| | |
|---|---|
| *LCCN* | 2020691656 |
| *Type of material* | Book |

# Bibliography

| | |
|---|---|
| *Main title* | Electromagnetic and acoustic wave tomography: direct and inverse problems in practical applications / edited by Nathan Blaunstein and Vladimir Yakubov. |
| *Published/Produced* | Boca Raton, FL.: CRC Press, Taylor and Francis Group, [2019] |
| *Description* | 1 online resource |
| *ISBN* | 9780429488276 ebook |
| | 9780429948404 ebook |
| | hb: acid-free paper |
| *LC classification* | TA1560 |
| *Related names* | Blaunstein, Nathan, editor. |
| | Yakubov, Vladimir, editor. |
| *Summary* | This book discusses the development of radio-wave tomography methods as a means of remote non-destructive testing, diagnostics of the internal structure of semi-transparent media, and reconstruction of the shapes of opaque objects based on multi-angle sounding. It describes physical-mathematical models of systems designed to reconstruct images of hidden objects, based on tomographic processing of multi-angle remote measurements of scattered radio and acoustic (ultrasonic) wave radiation-- Provided by publisher. |
| *LC Subjects* | Three-dimensional imaging--Mathematics. |
| | Tomography. |
| | Acoustic emission testing. |
| | Electromagnetic waves--Mathematical models. |
| | Remote sensing. |
| | Radar. |
| *Notes* | "A CRC title, part of the Taylor & Francis imprint, a member of the Taylor & Francis Group, the academic division of T&F Informa plc." |
| | Includes bibliographical references and index. |
| *Additional formats* | Print version: Electromagnetic and acoustic wave tomography Boca Raton: CRC Press, Taylor & Francis Group, [2019] 9781138490734 (hb: acid-free paper) (DLC) 2018006943 |

## Electromagnetic fields and waves: fundamentals of engineering

| | |
|---|---|
| *LCCN* | 2019946268 |
| *Type of material* | Book |
| *Personal name* | Riad, Sedki M. 1946- author. |
| *Main title* | Electromagnetic fields and waves: fundamentals of engineering / Sedki M. Riad, Imam M. Salama. |
| *Published/Produced* | New York: McGraw Hill, [2020] |
| *Description* | xxii, 666 pages: illustrations, maps; 27 cm |
| *ISBN* | 9781260457148 hardcover |
| | 1260457141 hardcover |
| *LC classification* | QC670 .R45 2020 |
| *Related names* | Salama, Imam M., author. |
| *LC Subjects* | Electromagnetic fields. |
| | Electromagnetic waves. |
| | Systems engineering. |
| *Other Subjects* | Electromagnetic fields. |
| | Electromagnetic waves. |
| | Systems engineering. |
| *Notes* | Includes index. |

## Electromagnetic metasurfaces: theory and applications

| | |
|---|---|
| *LCCN* | 2021010978 |
| *Type of material* | Book |
| *Personal name* | Achouri, Karim, author. |
| *Main title* | Electromagnetic metasurfaces: theory and applications / Karim Achouri, Christophe Caloz. |
| *Published/Produced* | Hoboken, NJ: Wiley-IEEE Press, 2021. |
| *Description* | 1 online resource |
| *ISBN* | 9781119525172 (epub) |
| | 9781119525196 (adobe pdf) |
| | (hardback) |
| *LC classification* | TK7871.15.M48 |
| *Related names* | Caloz, Christophe, 1969- author. |
| *Summary* | "This book introduces fundamental principles as well as applications of metasurfaces, i.e., electromagnetically thin structures manipulating EM wave propagation. The authors describe the precursors and history of metasurfaces before moving on to explore the physical insights that can be gained |

|               |                                                                 |
| ------------- | --------------------------------------------------------------- |
|               | from the material parameters of the metasurface. They also present how to compute the fields scattered by a metasurface, with known material parameters, being illuminated by an arbitrary incident field, as well as how to realize a practical metasurface and relate it its material parameters to physical structures. The book finishes with a discussion of the future of the field"-- Provided by publisher. |
| *Contents*    | Electromagnetic properties of materials - Metasurface modeling - Susceptibility synthesis - Scattered field computation - Practical implementation - Applications. |
| *LC Subjects* | Metasurfaces. Electromagnetic waves--Scattering.                |
| *Notes*       | Includes bibliographical references and index.                  |
| *Additional formats* | Print version: Achouri, Karim. Electromagnetic metasurfaces Hoboken, NJ: Wiley-IEEE Press, 2021 9781119525165 (DLC) 2021010977 |

## Electromagnetic metasurfaces: theory and applications

|                        |                                                            |
| ---------------------- | ---------------------------------------------------------- |
| *LCCN*                 | 2021010977                                                 |
| *Type of material*     | Book                                                       |
| *Personal name*        | Achouri, Karim, author.                                    |
| *Main title*           | Electromagnetic metasurfaces: theory and applications / Karim Achouri, Christophe Caloz. |
| *Published/Produced*   | Hoboken, NJ: Wiley-IEEE Press, 2021.                       |
| *ISBN*                 | 9781119525165 (hardback) (adobe pdf) (epub)                |
| *LC classification*    | TK7871.15.M48 A25 2021                                     |
| *Related names*        | Caloz, Christophe, 1969- author.                           |
| *Summary*              | "This book introduces fundamental principles as well as applications of metasurfaces, i.e., electromagnetically thin structures manipulating EM wave propagation. The authors describe the precursors and history of metasurfaces before moving on to explore the physical insights that can be gained from the material parameters of the metasurface. They also present how to compute the fields scattered |

|   |   |
|---|---|
| | by a metasurface, with known material parameters, being illuminated by an arbitrary incident field, as well as how to realize a practical metasurface and relate it its material parameters to physical structures.The book finishes with a discussion of the future of the field"-- Provided by publisher. |
| Contents | Electromagnetic properties of materials - Metasurface modeling - Susceptibility synthesis - Scattered field computation - Practical implementation - Applications. |
| LC Subjects | Metasurfaces.<br>Electromagnetic waves--Scattering. |
| Notes | Includes bibliographical references and index. |
| Additional formats | Online version: Achouri, Karim. Electromagnetic metasurfaces Hoboken, NJ: Wiley-IEEE Press, 2021 9781119525196 (DLC) 2021010978 |

**Electromagnetic radiation: history, theory and research**

|   |   |
|---|---|
| LCCN | 2020685739 |
| Type of material | Book |
| Main title | Electromagnetic radiation: history, theory and research / Constantinos Koutsojannis, editor. |
| Published/Produced | New York: Nova Science Publishers, [2018] |
| Description | 1 online resource |
| ISBN | 9781536143324 ebook<br>paperback |
| LC classification | QC661 |
| Related names | Koutsojannis, Constantinos, editor. |
| LC Subjects | Electromagnetic waves. |
| Notes | Includes bibiliographical references and index. |
| Additional formats | Print version: Electromagnetic radiation New York: Nova Science Publishers, [2018] 9781536143317 (paperback) (DLC) 2018275538 |
| Series | Physics research and technology |

**Electromagnetic radiation: history, theory and research**

|   |   |
|---|---|
| LCCN | 2018275538 |
| Type of material | Book |

| | |
|---|---|
| *Main title* | Electromagnetic radiation: history, theory and research / Constantinos Koutsojannis, editor. |
| *Published/Produced* | New York: Nova Science Publishers, [2018] |
| *Description* | x, 166 pages: illustrations; 23 cm. |
| *ISBN* | 9781536143317 (paperback) |
| | ebook |
| *LC classification* | QC661 .E385 2018 |
| *Related names* | Koutsojannis, Constantinos, editor. |
| *LC Subjects* | Electromagnetic waves. |
| *Notes* | Includes bibiliographical references and index. |
| *Series* | Physics research and technology |

**Electromagnetic radiation, scattering, and diffraction**

| | |
|---|---|
| *LCCN* | 2021030663 |
| *Type of material* | Book |
| *Personal name* | Pathak, P. H. (Prabhakar Harihar), 1942- author. |
| *Main title* | Electromagnetic radiation, scattering, and diffraction / Prabhakar H. Pathak and Robert J. Burkholder. |
| *Edition* | First edition. |
| *Published/Produced* | Hoboken: Wiley-IEEE Press, [2021] |
| *Description* | 1 online resource |
| *ISBN* | 9781119810537 (epub) |
| | 9781119810520 (adobe pdf) |
| | (cloth) |
| *LC classification* | QC661 |
| *Related names* | Burkholder, R. J. (Robert James), author. |
| *Summary* | "This book is designed to provide an understanding of the behavior of EM fields in radiation, scattering and guided wave environments, from first principles and from low to high frequencies. Physical interpretations of the EM wave phenomena are stressed along with their underlying mathematics. Fundamental principles are stressed, and numerous examples are included to illustrate concepts. This book can facilitate students with a somewhat limited undergraduate EM background to rapidly and systematically advance their understanding of EM wave theory that is useful and important for doing graduate level research on wave EM problems. This |

|  |  |
|---|---|
|  | book can therefore also be useful for gaining a better understanding of problems they are trying to simulate with commercial EM software and how to better interpret their results. The book can also be used for self-study as a refresher for EM industry professionals"-- Provided by publisher. |
| LC Subjects | Electromagnetic waves. |
|  | Electromagnetic waves--Scattering. |
|  | Electromagnetic waves--Diffraction. |
| Notes | Includes bibliographical references and index. |
| Additional formats | Print version: Pathak, P. H. 1942- Electromagnetic radiation, scattering, and diffraction First edition. Hoboken: Wiley-IEEE Press, [2021] 9781119810513 (DLC) 2021030662 |
| Series | IEEE Press series on electromagnetic wave theory |

**Electromagnetic radiation, scattering, and diffraction**

|  |  |
|---|---|
| LCCN | 2021030662 |
| Type of material | Book |
| Personal name | Pathak, P. H. (Prabhakar Harihar), 1942- author. |
| Main title | Electromagnetic radiation, scattering, and diffraction / Prabhakar H. Pathak and Robert J. Burkholder. |
| Edition | First edition. |
| Published/Produced | Hoboken: Wiley-IEEE Press, [2021] |
| ISBN | 9781119810513 (cloth) |
|  | (adobe pdf) |
|  | (epub) |
| LC classification | QC661 .P38 2021 |
| Related names | Burkholder, R. J. (Robert James), author. |
| Summary | "This book is designed to provide an understanding of the behavior of EM fields in radiation, scattering and guided wave environments, from first principles and from low to high frequencies. Physical interpretations of the EM wave phenomena are stressed along with their underlying mathematics. Fundamental principles are stressed, and numerous examples are included to illustrate concepts. This book can facilitate students with a somewhat limited undergraduate EM background to rapidly and |

|  |  |
|---|---|
|  | systematically advance their understanding of EM wave theory that is useful and important for doing graduate level research on wave EM problems. This book can therefore also be useful for gaining a better understanding of problems they are trying to simulate with commercial EM software and how to better interpret their results. The book can also be used for self-study as a refresher for EM industry professionals"-- Provided by publisher. |
| *LC Subjects* | Electromagnetic waves.<br>Electromagnetic waves--Scattering.<br>Electromagnetic waves--Diffraction. |
| *Notes* | Includes bibliographical references and index. |
| *Additional formats* | Online version: Pathak, Prabhakar H. Electromagnetic radiation, scattering, and diffraction First edition. Hoboken: Wiley, 2021 9781119810520 (DLC) 2021030663 |
| *Series* | IEEE Press series on electromagnetic wave theory |

**Electromagnetic radiation**

|  |  |
|---|---|
| *LCCN* | 2018953425 |
| *Type of material* | Book |
| *Personal name* | Freeman, R. R. (Richard R.), author. |
| *Main title* | Electromagnetic radiation / Richard Freeman, James King, Gregory Lafyatis. |
| *Edition* | First edition. |
| *Published/Produced* | Oxford, United Kingdom: Oxford University Press, 2019. |
| *Description* | xii, 624 pages: illustrations; 25 cm |
| *ISBN* | 0198726503<br>9780198726500 |
| *LC classification* | QC661 .F89 2019 |
| *Related names* | King, James (Senior scientist), author.<br>Lafyatis, Gregory P., author. |
| *Summary* | This graduate level textbook aims to teach fundamental ideas of advanced classical electrodynamics, with an emphasis on the physics of radiation. The text describes concepts with the minimum required mathematical detail, while the |

|  |  |
|---|---|
| Contents | accompanying side notes and end of chapter discussions provide the detailed derivations. Part I: Introductory Foundations - 1: Essentials of Electricity and Magnetism - 2: The Potentials - Part II: Origins of Radiation Fields - 3: General Relations between Fields and Sources - 4: Fields in terms of the Multiple Moments of the Source - Part III: Electromagnetism and Special Relativity - 5: Introduction to Special Relativity - 6: Radiation from Charges Moving at Relativistic Velocities - 7: Relativistic Electrodynamics - 8: Field Reactions to Moving Charges - Part IV: Radiation in Materials - 9: Properties of Electromagnetic Radiation in Materials - 10: Models of Electromagnetic Response of Materials - 11: Scattering of Electromagnetic Radiation in Materials - 12: Diffraction and the Propagation of Light - 13: Radiation Fields in Constrained Environments. |
| LC Subjects | Electromagnetic waves. |
| Other Subjects | Electromagnetic waves. |
| Notes | Includes bibliographical references and index. |
| Additional formats | Ebook version: Freeman, Richard R. Electromagnetic Radiation. Oxford: Oxford University Press USA - OSO, ©2019 9780191039973 |

**Electromagnetic scattering: a remote sensing perspective**

|  |  |
|---|---|
| LCCN | 2016059831 |
| Type of material | Book |
| Main title | Electromagnetic scattering: a remote sensing perspective / Yang Du, Zhejiang University, China. |
| Published/Produced | New Jersey: World Scientific, [2017] |
| Description | xii, 397 pages: illustrations (some color); 24 cm |
| ISBN | 9789813209862 (hardback; alk. paper) |
|  | 9813209860 (hardback; alk. paper) |
| LC classification | QC665.S3 E435 2017 |
| Related names | Du, Yang (Professor of electrical engineering), editor. |
| Summary | "Remote sensing is a fast-growing field with many important applications as demonstrated in the |

numerous scientific missions of the Earth Observation System (EOS) worldwide. Given the inter-disciplinary nature of remote sensing technologies, the fulfillment of these scientific goals calls for, among other things, a fundamental understanding of the complex interaction between electromagnetic waves and the targets of interest. Using a systematic treatment, Electromagnetic Scattering: A Remote Sensing Perspective presents some of the recently advanced methods in electromagnetic scattering, as well as updates on the current progress on several important aspects of such an interaction. The book covers topics including scattering from random rough surfaces of both terranean and oceanic natures, scattering from typical man-made targets or important canonical constituents of natural scenes, such as a dielectric finite cylinder or dielectric thin disk, the characterization of a natural scene as a whole represented as a random medium, and the extraction of target features with a polarimetric radar"-- Provided by publisher.

| | |
|---|---|
| *LC Subjects* | Electromagnetic waves--Scattering. |
| | Remote sensing. |
| *Notes* | Includes bibliographical references and index. |

**Electromagnetic vortices: wave phenomena and engineering applications**

| | |
|---|---|
| *LCCN* | 2021030661 |
| *Type of material* | Book |
| *Main title* | Electromagnetic vortices: wave phenomena and engineering applications / edited by Zhi Hao Jiang, Douglas H. Werner. |
| *Published/Produced* | Hoboken, New Jersey: Wiley-IEEE Press, [2021] |
| *Description* | 1 online resource |
| *ISBN* | 9781119662877 (epub) |
| | 9781119662969 (adobe pdf) |
| | (cloth) |
| *LC classification* | QC718.5.E45 |
| *Related names* | Jiang, Zhi Hao, editor. |

Werner, Douglas H., 1960- editor.

*Summary* "This book describes cutting-edge research and development on electromagnetic vortex waves, including their related wave properties and a variety of potential transformative applications. It is divided into three parts. The first part presents the generation/sorting/manipulation and interesting wave behavior of vortex waves in the infrared and optical regimes with custom-designed nanostructures. The second cluster deals with the generation/ multiplexing/propagation of vortex waves at microwave and millimeter-wave frequencies. The last group of chapters exhibit several representative practical applications of vortex waves from a system perspective ranging from microwave frequencies to optical wavelengths. Each chapter is contributed by an internationally recognized author or group of authors covering the latest research results"-- Provided by publisher.

*Contents* Fundamentals of orbital angular momentum beams: concepts, antenna analogies, and applications / Anastasios Papathanasopoulos, Yahya Rahmat-Samii - OAM radio-physical foundations and applications of electromagnetic orbital angular momentum in radio science and technology / Bo Thidé, Fabrizio Tamburini - Generation of microwave vortex beams using metasurfaces / Jia Yuan Yin, Tie Jun Cui - Application of transformation optics and 3D printing technology in vortex wave generation / Jianjia Yi, Shah Nawaz Burokur, Douglas H. Werner - Millimeter-wave transmit-arrays for high-capacity and wideband generation of scalar and vector vortex beams / Zhi Hao Jiang, Lei Kang, Wei Hong, Douglas H. Werner - Twisted light with metamaterials / Natalia Litchinitser - Generation of optical vortex beams / Yuanjie Yang, Cheng-Wei Qiu - Orbital angular momentum generation, detection and angular momentum conservation with second harmonic

|  |  |
|---|---|
|  | generation / Menglin L.N. Chen, Xiaoyan Y.Z. Xiong, Wei E.I. Sha, Li Jun Jiang - Orbital angular momentum based structured radio beams and its applications / Xianmin Zhang, Shilie Zheng, Wei E.I. Sha, Li Jun Jiang, Xiaowen Xiong, Zelin Zhu, Zhixia Wang, Yuqi Chen, Jiayu Zheng, Xinyue Wang, Menglin L.N. Chen - OAM multiplexing using uniform circular array and microwave circuit for short-range communication / Kentaro Murata, Naoki Honma - OAM communications in multipath environments / Xiaoming Chen, Wei Xue - High-capacity free-space communications using multiplexing of multiple OAM beams / Alan Willner, Runzhou Zhang, Kai Pang, Haoqian Song, Cong Liu, Hao Song, Xinzhou Su, Huibin Zhou, Nanzhou Hu, Zhe Zhao, Guodong Xie, Yongxiong Ren, Hao Huang, Moshe Tur - Theory of vector beams for chirality and magnetism detection of subwavelength particles / Filippo Capolino, Mina Hanifeh - Quantum applications of structured photons / Alessio D'Errico, Ebrahim Karimi. |
| *LC Subjects* | Electromagnetic waves.<br>Spheromaks.<br>Angular momentum.<br>Nanostructures. |
| *Notes* | Includes bibliographical references and index. |
| *Additional formats* | Print version: Electromagnetic vortices Hoboken, New Jersey: Wiley-IEEE Press, [2021] 9781119662822 (DLC) 2021030660 |

**Electromagnetic vortices: wave phenomena and engineering applications**

|  |  |
|---|---|
| *LCCN* | 2021030660 |
| *Type of material* | Book |
| *Main title* | Electromagnetic vortices: wave phenomena and engineering applications / edited by Zhi Hao Jiang, Douglas H. Werner. |
| *Published/Produced* | Hoboken, New Jersey: Wiley-IEEE Press, [2021] |
| *ISBN* | 9781119662822 (cloth) |

|  |  |
|---|---|
| | (adobe pdf) |
| | (epub) |
| LC classification | QC718.5.E45 .E44 2021 |
| Related names | Jiang, Zhi Hao, editor. |
| | Werner, Douglas H., 1960- editor. |
| Summary | "This book describes cutting-edge research and development on electromagnetic vortex waves, including their related wave properties and a variety of potential transformative applications. It is divided into three parts. The first part presents the generation/sorting/manipulation and interesting wave behavior of vortex waves in the infrared and optical regimes with custom-designed nanostructures. The second cluster deals with the generation/multiplexing/propagation of vortex waves at microwave and millimeter-wave frequencies. The last group of chapters exhibit several representative practical applications of vortex waves from a system perspective ranging from microwave frequencies to optical wavelengths. Each chapter is contributed by an internationally recognized author or group of authors covering the latest research results"-- Provided by publisher. |
| Contents | Fundamentals of orbital angular momentum beams: concepts, antenna analogies, and applications / Anastasios Papathanasopoulos, Yahya Rahmat-Samii - OAM radio-physical foundations and applications of electromagnetic orbital angular momentum in radio science and technology / Bo Thidé, Fabrizio Tamburini - Generation of microwave vortex beams using metasurfaces / Jia Yuan Yin, Tie Jun Cui - Application of transformation optics and 3D printing technology in vortex wave generation / Jianjia Yi, Shah Nawaz Burokur, Douglas H. Werner - Millimeter-wave transmit-arrays for high-capacity and wideband generation of scalar and vector vortex beams / Zhi Hao Jiang, Lei Kang, Wei Hong, Douglas H. Werner - Twisted light with metamaterials / Natalia |

|  |  |
|---|---|
|  | Litchinitser - Generation of optical vortex beams / Yuanjie Yang, Cheng-Wei Qiu - Orbital angular momentum generation, detection and angular momentum conservation with second harmonic generation / Menglin L.N. Chen, Xiaoyan Y.Z. Xiong, Wei E.I. Sha, Li Jun Jiang - Orbital angular momentum based structured radio beams and its applications / Xianmin Zhang, Shilie Zheng, Wei E.I. Sha, Li Jun Jiang, Xiaowen Xiong, Zelin Zhu, Zhixia Wang, Yuqi Chen, Jiayu Zheng, Xinyue Wang, Menglin L.N. Chen - OAM multiplexing using uniform circular array and microwave circuit for short-range communication / Kentaro Murata, Naoki Honma - OAM communications in multipath environments / Xiaoming Chen, Wei Xue - High-capacity free-space communications using multiplexing of multiple OAM beams / Alan Willner, Runzhou Zhang, Kai Pang, Haoqian Song, Cong Liu, Hao Song, Xinzhou Su, Huibin Zhou, Nanzhou Hu, Zhe Zhao, Guodong Xie, Yongxiong Ren, Hao Huang, Moshe Tur - Theory of vector beams for chirality and magnetism detection of subwavelength particles / Filippo Capolino, Mina Hanifeh - Quantum applications of structured photons / Alessio D'Errico, Ebrahim Karimi. |
| *LC Subjects* | Electromagnetic waves.<br>Spheromaks.<br>Angular momentum.<br>Nanostructures. |
| *Notes* | Includes bibliographical references and index. |
| *Additional formats* | Online version: Electromagnetic vortices Hoboken, New Jersey: Wiley-IEEE Press, [2021] 9781119662969 (DLC) 2021030661 |

**Electromagnetic wave absorbers: detailed theories and applications**

|  |  |
|---|---|
| *LCCN* | 2019003710 |
| *Type of material* | Book |
| *Personal name* | Kotsuka, Y. (Youji), 1941- author. |

| | |
|---|---|
| *Main title* | Electromagnetic wave absorbers: detailed theories and applications / Youji Kotsuka. |
| *Published/Produced* | Hoboken, New Jersey: John Wiley & Sons, Inc., 2019. <br> ©2019 |
| *ISBN* | 9781119564126 (hardback) <br> 1119564123 (hardback) <br> (ePDF) <br> (ePDF) <br> (epub) <br> (epub) |
| *LC classification* | QC665.T7 K68 2019 |
| *Summary* | "Provides detailed explanations of basic theory and applied theory for understanding EM-wave absorbers -Discusses the material constant measurement methods of EM-wave absorption characteristics that are necessary for designing the EM-wave absorber - Includes examples of EM-wave absorber configurations In recent years, along with increases in various kinds of demands on communication systems, the communication environment has become more and more complicated. As a result, the introduction of EM-wave absorbers is required in each technical field, especially from the viewpoint of preventing interference and mutual interaction. Demands for various EM-wave absorbers are rapidly increasing along with the recent trends towards complicated electromagnetic environments and development of higher-frequency communication equipment. Further, the rapid progress of recent artificial intelligence (AI) technology is now remarkable. With its rapid progress the need of electrical controllable EM-wave absorbers is being realized"-- Provided by publisher. |
| *Contents* | Fundamentals of electromagnetic wave absorbers - Fundamental theory of EM-wave absorber - Methods of absorber analysis - Basic theory of computer analysis - Fundamental EM-wave absorber materials - Theory of special mediums - Measurement methods |

|                     | on EM-wave absorbers - Configuration examples of the EM-wave absorber - Absorber characteristics control by equivalent transformation method of material constants - Autonomous controllable absorber. |
|---|---|
| *LC Subjects*       | Electromagnetic waves--Transmission. Absorption. |
| *Notes*             | "Published simultaneously in Canada"--Title page verso. |
|                     | Includes bibliographical references and index. |
| *Additional formats*| Online version: Kotsuka, Y. (Youji), 1941- author. Electromagnetic wave absorbers Hoboken, New Jersey: John Wiley & Sons, Inc., 2019 9781119564140 (DLC) 2019011758 |

**Electromagnetic wave absorbers: detailed theories and applications**

| | |
|---|---|
| *LCCN* | 2019011758 |
| *Type of material* | Book |
| *Personal name* | Kotsuka, Y. (Youji), 1941- author. |
| *Main title* | Electromagnetic wave absorbers: detailed theories and applications / Youji Kotsuka. |
| *Published/Produced* | Hoboken, New Jersey: John Wiley & Sons, Inc., 2019. |
| | ©2019 |
| *Description* | 1 online resource. |
| *ISBN* | 9781119564140 (Adobe PDF) |
| | 9781119564386 (ePub) |
| | 1119564387q(epub) |
| | 9781119564386q(epub) |
| | 9781119564140q(ePDF) |
| *LC classification* | QC665.T7 |
| *Summary* | "Provides detailed explanations of basic theory and applied theory for understanding EM-wave absorbers -Discusses the material constant measurement methods of EM-wave absorption characteristics that are necessary for designing the EM-wave absorber - Includes examples of EM-wave absorber configurations In recent years, along with increases in various kinds of demands on communication |

|  |  |
|---|---|
|  | systems, the communication environment has become more and more complicated. As a result, the introduction of EM-wave absorbers is required in each technical field, especially from the viewpoint of preventing interference and mutual interaction. Demands for various EM-wave absorbers are rapidly increasing along with the recent trends towards complicated electromagnetic environments and development of higher-frequency communication equipment. Further, the rapid progress of recent artificial intelligence (AI) technology is now remarkable. With its rapid progress the need of electrical controllable EM-wave absorbers is being realized"-- Provided by publisher. |
| Contents | Fundamentals of electromagnetic wave absorbers - Fundamental theory of EM-wave absorber - Methods of absorber analysis - Basic theory of computer analysis - Fundamental EM-wave absorber materials - Theory of special mediums - Measurement methods on EM-wave absorbers - Configuration examples of the EM-wave absorber - Absorber characteristics control by equivalent transformation method of material constants - Autonomous controllable absorber. |
| LC Subjects | Electromagnetic waves--Transmission. Absorption. |
| Notes | "Published simultaneously in Canada"--Title page verso. Includes bibliographical references and index. |
| Additional formats | Print version: Kotsuka, Y. (Youji), 1941- author. Electromagnetic wave absorbers Hoboken, New Jersey: John Wiley & Sons, Inc., 2019 9781119564126 (DLC) 2019003710 |

**Electromagnetic wave propagation, radiation, and scattering: from fundamentals to applications**

|  |  |
|---|---|
| LCCN | 2017303173 |
| Type of material | Book |
| Personal name | Ishimaru, Akira, 1928- author. |

| | |
|---|---|
| *Main title* | Electromagnetic wave propagation, radiation, and scattering: from fundamentals to applications / Akira Ishimaru, University of Washington, Seattle, WA, USA; IEEE Antennas and Propagation Society, sponsor. |
| *Edition* | Second edition. |
| *Published/Produced* | Piscataway, NJ: IEEE Press/Wiley, [2017] |
| *Description* | xxvii, 937 pages: illustrations; 24 cm. |
| *ISBN* | 9781118098813 (hardcover) |
| *LC classification* | QC665.T7 I74 2017 |
| *LC Subjects* | Electromagnetic waves--Transmission. Electromagnetic waves--Scattering. |
| *Notes* | Includes bibliographical references (pages 913-928) and index. |
| *Series* | The IEEE Press series on electromagnetic wave theory |

## Electromagnetic waves, materials, and computation with MATLAB

| | |
|---|---|
| *LCCN* | 2017012371 |
| *Type of material* | Book |
| *Personal name* | Kalluri, Dikshitulu K., author. |
| *Main title* | Electromagnetic waves, materials, and computation with MATLAB / Dikshitulu K. Kalluri. |
| *Edition* | Second edition. |
| *Published/Produced* | Boca Raton: Taylor & Francis, CRC Press, 2017. |
| *ISBN* | 9781498733403 (hardback: alk. paper) |
| *LC classification* | QC760 .K35 2017 |
| *Incomplete contents* | v. 1. Advanced electromagnetic computation |
| *LC Subjects* | MATLAB. Electromagnetic waves--Computer simulation. Materials--Electric properties. |
| *Notes* | Includes bibliographical references. |

## EMF effects from power sources and electrosmog

| | |
|---|---|
| *LCCN* | 2018047175 |
| *Type of material* | Book |
| *Personal name* | Rea, William J. (Physician), author. |
| *Main title* | EMF effects from power sources and electrosmog / William J. Rea. |

| | |
|---|---|
| *Published/Produced* | Boca Raton: CRC Press, Taylor & Francis Group, [2019] |
| *Description* | xiii, 149 pages: illustrations; 22 cm. |
| *ISBN* | 9780367031091 (hardback: acid-free paper) |
| | ebook |
| *LC classification* | RA569.3 R43 2019 |
| *LC Subjects* | Electromagnetic fields--Health aspects. |
| | Electromagnetic waves--Health aspects. |
| | Electric lines--Health aspects. |
| | Stray currents--Physiological effect. |
| | Electromagnetism--Physiological effect. |
| *Notes* | Includes bibliographical references and index. |
| *Series* | CRC focus series |
| | Electromagnetic frequency sensitivities series |

**EMF effects from power sources and electrosmog**

| | |
|---|---|
| *LCCN* | 2020693175 |
| *Type of material* | Book |
| *Personal name* | Rea, William J. (Physician), author. |
| *Main title* | EMF effects from power sources and electrosmog / William J. Rea. |
| *Published/Produced* | Boca Raton: CRC Press, Taylor and Francis Group, [2019] |
| *Description* | 1 online resource |
| *ISBN* | 9780429670336 epub |
| | 9780429020520 ebook |
| | hardback: acid-free paper |
| *LC classification* | RA569.3 |
| *LC Subjects* | Electromagnetic fields--Health aspects. |
| | Electromagnetic waves--Health aspects. |
| | Electric lines--Health aspects. |
| | Stray currents--Physiological effect. |
| | Electromagnetism--Physiological effect. |
| *Notes* | Includes bibliographical references and index. |
| *Additional formats* | Print version: EMF effects from power sources and electrosmog Boca Raton: CRC Press, Taylor & Francis Group, [2019] 9780367031091 (hardback: acid-free paper) (DLC) 2018047175 |
| *Series* | CRC focus series |

Electromagnetic frequency sensitivities series

**Extreme events in geospace: origins, predictability, and consequences**

| | |
|---|---|
| LCCN | 2018302770 |
| Type of material | Book |
| Main title | Extreme events in geospace: origins, predictability, and consequences / edited by Natalia Buzulukova. |
| Published/Produced | Amsterdam: Elsevier, [2018] ©2018 |
| Description | liii, 744 pages: illustrations (some color); 24 cm |
| ISBN | 9780128127001 (pbk.) 0128127007 (pbk.) |
| LC classification | QB505 .E98 2018 |
| Summary | Extreme Events in Geospace: Origins, Predictability, and Consequences helps deepen the understanding, description, and forecasting of the complex and inter-related phenomena of extreme space weather events. Composed of chapters written by representatives from many different institutions and fields of space research, the book offers discussions ranging from definitions and historical knowledge to operational issues and methods of analysis. Given that extremes in ionizing radiation, ionospheric irregularities, and geomagnetically induced currents may have the potential to disrupt our technologies or pose danger to human health, it is increasingly important to synthesize the information available on not only those consequences but also the origins and predictability of such events. Extreme Events in Geospace: Origins, Predictability, and Consequences is a valuable source for providing the latest research for geophysicists and space weather scientists, as well as industries impacted by space weather events, including GNSS satellites and radio communication, power grids, aviation, and human spaceflight.-- Source other than Library of Congress. |
| Contents | Part 1. Overview of impacts and effects - Linking space weather science to impacts - the view from the earth - Part 2. Solar origins and statistics of extremes |

- Extreme solar eruptions and their space weather consequences - Solar flare forecasting: present methods and challenges - Geoeffectiveness of solar and interplanetary structures and generation of strong geomagnetic storms - Statistics of extreme space weather events - Data-driven modeling of extreme space weather - Part 3. Geomagnetic storms and geomagnetically induced currents - Supergeomagnetic storms: past, present, and future An overview of science challenges pertaining to our understanding of extreme geomagnetically induced currents - Extreme-event geoelectric hazard maps - Geomagnetic storms: first-principles models for extreme geospace environment - Empirical modeling of extreme events: storm-time geomagnetic field, electric current, and pressure distributions - Part 4. Plasma and radiation environment - Extreme space weather events: a GOES perspective - Near-Earth radiation environment for extreme solar and geomagnetic conditions - Magnetospheric "killer" relativistic electron dropouts (REDs) and repopulation: a cyclical process - Extreme space weather spacecraft surface charging and arcing effects - Deep dielectric charging and spacecraft anomalies - Solar particle events and human deep space exploration: measurements and considerations - Characterizing the variation in atmospheric radiation at aviation altitudes - High-energy transient luminous atmospheric phenomena: the potential danger for suborbital flights - Part 5. Ionospheric/thermospheric effects and impacts - Ionosphere and thermosphere responses to extreme geomagnetic storms - How might the thermosphere and ionosphere react to an extreme space weather event? - The effect of solar radio bursts on GNSS signals - Extreme ionospheric storms and their effects on GPS systems - Recent geoeffective space weather events and technological system impacts - Extreme space weather in time: effects on Earth - Part 6.

|  |  |
|---|---|
| | Dealing with the space weather - Dealing with space weather: the Canadian experience - Space weather: what are policymakers seeking? - Extreme space weather and emergency management - The social and economic impacts of moderate and severe space weather - Severe space weather events in the Australian context - Extreme space weather research in Japan. |
| *LC Subjects* | Space environment. |
| | Electromagnetic waves. |
| | Solar energetic particles. |
| | Solar wind. |
| | Magnetic storms. |
| *Other Subjects* | Electromagnetic waves. |
| | Magnetic storms. |
| | Solar energetic particles. |
| | Solar wind. |
| | Space environment. |
| *Notes* | Includes bibliographical references and index. |

**From ER to E.T.: how electromagnetic technologies are changing our lives**

|  |  |
|---|---|
| *LCCN* | 2017446414 |
| *Type of material* | Book |
| *Personal name* | Bansal, Rajeev, author. |
| *Main title* | From ER to E.T.: how electromagnetic technologies are changing our lives / Rajeev Bansal. |
| *Published/Produced* | Piscataway, NJ: IEEE Press; Hoboken, New Jersey: Wiley, [2017] |
| *Description* | xii, 207 pages: illustrations; 24 cm. |
| *ISBN* | 9781118458174 paperback |
| | 1118458176 paperback |
| *LC classification* | QC661 .B253 2017 |
| *LC Subjects* | Electromagnetic waves--Technology. |
| | Electromagnetic fields. |
| *Notes* | Includes bibliographical references and index. |
| *Series* | IEEE Press series on electromagnetic wave theory |
| | IEEE Press series on electromagnetic wave theory. |

**Gravitational wave astrophysics: early results from gravitational wave searches and electromagnetic counterparts: proceedings of the 338th symposium of the International Astronomical Union held in Baton Rouge, United States, October 16-19, 2017**

| | |
|---|---|
| *LCCN* | 2018418571 |
| *Type of material* | Book |
| *Corporate name* | International Astronomical Union. Symposium (338th: 2017: Baton Rouge, Louisiana, United States). |
| *Main title* | Gravitational wave astrophysics: early results from gravitational wave searches and electromagnetic counterparts: proceedings of the 338th symposium of the International Astronomical Union held in Baton Rouge, United States, October 16-19, 2017 / edited by Gabriela González and Robert Hynes. |
| *Published/Produced* | Cambridge: Cambridge University Press, 2019. |
| *Description* | x, 107 pages: illustrations; 26 cm |
| *Links* | https://www.cambridge.org/core/journals/proceedings-of-the-international-astronomical-union/issue/1F039DD2202C2DB96A5E23490609C95A |
| *ISBN* | 9781107192591 (hbk.) |
| | 1107192595 |
| *LC classification* | QB651 .I57 2017 |
| *Variant title* | IAU 338 |
| *Related names* | González, Gabriela, 1965- editor. |
| | Hynes, Robert J. (Robert Joseph), 1937- editor. |
| *LC Subjects* | LIGO (Observatory) |
| | Gravitational waves--Congresses. |
| | Astrophysics--Congresses. |
| | Electromagnetic fields--Congresses. |
| *Other Subjects* | Astrophysics. |
| | Electromagnetic fields. |
| | Gravitational waves. |
| *Form/Genre* | Conference papers and proceedings. |
| *Notes* | Includes bibliographical references and index. |
| *Additional formats* | Online version: International Astronomical Union. Symposium. Gravitational wave astrophysics (IAU S338). Cambridge: Cambridge University Press, [2018] (OCoLC)1091628792 |

| | |
|---|---|
| *Series* | IAU symposium proceedings series. 1743-9213 |
| | IAU symposium and colloquium proceedings series. 1743-9213 |

## Handbook of radar scattering statistics for terrain

| | |
|---|---|
| *LCCN* | 2019286507 |
| *Type of material* | Book |
| *Personal name* | Ulaby, Fawwaz T. (Fawwaz Tayssir), 1943- author. |
| *Main title* | Handbook of radar scattering statistics for terrain / Fawwaz T. Ulaby, The Radiation Laboratory, M. Craig Dobson, The University of Michigan; with updated Python and MATLAB© software by José Luis Álvarez-Pérez. |
| *Published/Produced* | Boston: Artech House, [2019] |
| *Description* | xiv, 403 pages: illustrations; 24 cm. |
| *ISBN* | 9781630817015 hardcover |
| | 1630817015 hardcover |
| *LC classification* | QC665.S3 U43 2019 |
| *Related names* | Dobson, M. Craig, author. |
| | Álvarez-Pérez, José Luis, author |
| *LC Subjects* | Electromagnetic waves--Scattering. |
| | Electromagnetic waves--Scattering--Tables. |
| *Other Subjects* | Electromagnetic waves--Scattering. |
| *Form/Genre* | Tables. |
| *Notes* | "First published in 1989, this update features a new preface, along with the new appendixes...This update combines the book and software, previously sold separately, into a single new resource." --Back cover |
| | Includes bibliographical references and index. |
| *Series* | The Artech House Remote Sensing Library |

## Harmonic modeling of voltage source converters using simple numerical methods

| | |
|---|---|
| *LCCN* | 2021015329 |
| *Type of material* | Book |
| *Personal name* | Lian, Ryan Kuo-Lung, author. |
| *Main title* | Harmonic modeling of voltage source converters using simple numerical methods / Ryan Kuo-Lung Lian, National Taiwan University of Science and |

|  |  |
|---|---|
|  | Technology, Taipei, Taiwan, Ramadhani Kurniawan Subroto, University of Brawijaya, Malang, Indonesia, Bing Hao Lin, National Taiwan University of Science and Technology, Taipei, Taiwan. |
| Published/Produced | Hoboken, NJ, USA: Wiley, 2021. |
| Description | 1 online resource |
| ISBN | 9781119527152 (epub) |
|  | 9781119527145 (adobe pdf) |
|  | (hardback) |
| LC classification | TK2799 |
| Related names | Subroto, Ramadhani Kurniawan, author. |
|  | Lin, Bing Hao, author. |
| Summary | "The ac electric power systems are essentially designed to operate with sinusoidal voltages and currents at frequencies of 50 or 60 Hz. However, certain types of power components or loads produce currents and voltages with frequencies that are integer multiples of these frequencies (i.e., the fundamental frequencies). These higher frequencies are a form of electrical pollution known as power system harmonics. Power system harmonics are not a new phenomenon, and it is as old as the distribution of alternating current, which began in 1895-1896 [5]. It is reported that in 1893, Charles Proteus Steinmetz had worked on the problem of motor heating while working at Thomson-Houston [6]. After rigorous calculations and experimental validation, Steinmetz concluded that the problem was due to the resonance in the transmission circuit feeding the plant and a generator with a substantial amount of waveform distortion. Consequently, Steinmetz proposed two solutions to overcome this harmonic problem. The first was to reduce the system frequency to one-half of its original value. That is, to reduce the original frequency value of 125 Hz to a new value of 62.5 Hz. Note that at that time, most of the single-phase generator were operated at 125 Hz, 140 Hz or 1331"-- Provided by publisher. |
| LC Subjects | Electric current converters--Mathematical models. |

|  |  |
|---|---|
| | Harmonics (Electric waves)--Mathematical models. |
| | Electromagnetic interference--Mathematical models. |
| | Numerical analysis. |
| | Electric power-plants--Equipment and supplies. |
| *Notes* | Includes bibliographical references and index. |
| *Additional formats* | Print version: Lian, Ryan Kuo-Lung. Harmonic modeling of voltage source converters using simple numerical methods Hoboken, NJ, USA: Wiley, 2021 9781119527138 (DLC) 2021015328 |

**Harmonic modeling of voltage source converters using simple numerical methods**

|  |  |
|---|---|
| *LCCN* | 2021015328 |
| *Type of material* | Book |
| *Personal name* | Lian, Ryan Kuo-Lung, author. |
| *Main title* | Harmonic modeling of voltage source converters using simple numerical methods / Ryan Kuo-Lung Lian, National Taiwan University of Science and Technology, Taipei, Taiwan, Ramadhani Kurniawan Subroto, University of Brawijaya, Malang, Indonesia, Bing Hao Lin, National Taiwan University of Science and Technology, Taipei, Taiwan. |
| *Published/Produced* | Hoboken, NJ, USA: Wiley, 2021. |
| *ISBN* | 9781119527138 (hardback) |
| | (adobe pdf) |
| | (epub) |
| *LC classification* | TK2799 .L53 2021 |
| *Related names* | Subroto, Ramadhani Kurniawan, author. |
| | Lin, Bing Hao, author. |
| *Summary* | "The ac electric power systems are essentially designed to operate with sinusoidal voltages and currents at frequencies of 50 or 60 Hz. However, certain types of power components or loads produce currents and voltages with frequencies that are integer multiples of these frequencies (i.e., the fundamental frequencies). These higher frequencies are a form of electrical pollution known as power system harmonics. Power system harmonics are not a new phenomenon, and it is as old as the distribution of |

|  |  |
|---|---|
|  | alternating current, which began in 1895-1896 [5]. It is reported that in 1893, Charles Proteus Steinmetz had worked on the problem of motor heating while working at Thomson-Houston [6]. After rigorous calculations and experimental validation, Steinmetz concluded that the problem was due to the resonance in the transmission circuit feeding the plant and a generator with a substantial amount of waveform distortion. Consequently, Steinmetz proposed two solutions to overcome this harmonic problem. The first was to reduce the system frequency to one-half of its original value. That is, to reduce the original frequency value of 125 Hz to a new value of 62.5 Hz. Note that at that time, most of the single-phase generator were operated at 125 Hz, 140 Hz or 1331"-- Provided by publisher. |
| *LC Subjects* | Electric current converters--Mathematical models. Harmonics (Electric waves)--Mathematical models. Electromagnetic interference--Mathematical models. Numerical analysis. Electric power-plants--Equipment and supplies. |
| *Notes* | Includes bibliographical references and index. |
| *Additional formats* | Online version: Lian, Ryan Kuo-Lung. Harmonic modeling of voltage source converters using simple numerical methods Hoboken, NJ, USA: Wiley, 2021 9781119527145 (DLC) 2021015329 |

## IEEE journal of electromagnetics, RF and microwaves in medicine and biology.

| | |
|---|---|
| *LCCN* | 2015203421 |
| *Type of material* | Periodical or Newspaper |
| *Main title* | IEEE journal of electromagnetics, RF and microwaves in medicine and biology. |
| *Published/Produced* | Piscataway, NJ: Institute of Electrical and Electronic Engineers, Inc., [2017-] |
| *Publication history* | Began with: Volume 1, Number 1 (June 2017). |
| *Current frequency* | Bimonthly |
| *ISSN* | 2469-7249 |
| *LC classification* | QC660.5 .I34 |

| | |
|---|---|
| *Variant title* | Journal of electromagnetics, RF and microwaves in medicine and biology |
| | Institute of Electrical and Electronics Engineers journal of electromagnetics, RF and microwaves in medicine and biology |
| | IJERLV |
| *Serial key title* | IEEE journal of electromagnetics, RF and microwaves in medicine and biology |
| *Abbreviated title* | IEEE j. electromagn. RF microw. med. biol. |
| *Related names* | Institute of Electrical and Electronics Engineers. |
| | IEEE Antennas and Propagation Society. |
| | IEEE Microwave Theory and Techniques Society. |
| | IEEE Engineering in Medicine and Biology Society. |
| | IEEE Sensors Council. |
| *LC Subjects* | Electromagnetic waves--Periodicals. |
| | Electromagnetism--Periodicals. |
| | Microwaves--Periodicals. |
| *Other Subjects* | Electromagnetic waves. |
| | Electromagnetism. |
| | Microwaves. |
| *Form/Genre* | Periodicals. |
| *Reproduction no./Source* | IEEE, 445 Hoes Ln., Piscataway, NJ 08854 |
| *Additional formats* | Online version: IEEE journal of electromagnetics, RF and microwaves in medicine and biology (Online) 2469-7257 (DLC) 2015203420 (OCoLC)930782004 |

**IEEE journal of microwaves.**

| | |
|---|---|
| *LCCN* | 2020202510 |
| *Type of material* | Periodical or Newspaper |
| *Main title* | IEEE journal of microwaves. |
| *Published/Produced* | Piscataway, NJ: Institute of Electrical and Electronics Engineers, Inc., 2021- |
| *Publication history* | Began with: Volume 1, issue 1 (winter 2021). |
| *Description* | Articles are published on an ongoing basis and organized in quarterly issues. |
| | Volume 1, issue 1 also called "Inaugural issue." |
| *Current frequency* | Quarterly |
| *ISSN* | 2692-8388 |

| | |
|---|---|
| *LC classification* | QC677 |
| *Variant title* | Journal of microwaves |
| | Institute of Electrical and Electronics Engineers journal of microwaves |
| | JMW |
| | IEEE j. microwaves |
| *Serial key title* | IEEE journal of microwaves |
| *Abbreviated title* | IEEE j. microw. |
| *Related names* | IEEE Microwave Theory and Techniques Society, issuing body. |
| *LC Subjects* | Microwaves--Periodicals. |
| | Electromagnetic waves--Periodicals. |
| *Other Subjects* | Electromagnetic waves. |
| | Microwaves. |
| *Form/Genre* | Periodicals. |
| *Notes* | "Expanding Science, Technology & Connectivity Around the Globe." |

**Introduction to electromagnetic waves with Maxwell's equations**

| | |
|---|---|
| *LCCN* | 2021014305 |
| *Type of material* | Book |
| *Personal name* | Ergül, Özgür, author. |
| *Main title* | Introduction to electromagnetic waves with Maxwell's equations / Ozgur Ergul. |
| *Published/Produced* | Hoboken, NJ: Wiley, 2021. |
| *Description* | 1 online resource |
| *ISBN* | 9781119626749 (epub) |
| | 9781119626732 (adobe pdf) |
| | (cloth) |
| *LC classification* | QC661 |
| *Summary* | "Written in an accessible style, and based on the author's extensive experience of teaching electromagnetics, this text "simply" presents the well-known topics that all electromagnetics instructors would include in their syllabus. The text is carefully tuned to be relevant to an audience of engineering students who have been exposed to the basic curriculum in linear algebra and multivariate calculus. Maxwell's equations form the backbone of |

|                     |                                                                                                                                                                                                                                                                                                                                                                                                                                                                             |
| ------------------- | --------------------------------------------------------------------------------------------------------------------------------------------------------------------------------------------------------------------------------------------------------------------------------------------------------------------------------------------------------------------------------------------------------------------------------------------------------------------------- |
|                     | the book; as they are developed step by step in consecutive chapters, related electromagnetic phenomena, as well as accompanying mathematical tools, are discussed simultaneously. Whilst the book will appeal to both undergraduate and post/graduate students, no prerequisite electromagnetics knowledge is required, as it starts with the most fundamental concepts. However, it is expected that the reader is familiar with basic calculus."-- Provided by publisher. |
| *LC Subjects*       | Electromagnetic waves.                                                                                                                                                                                                                                                                                                                                                                                                                                                      |
|                     | Maxwell equations.                                                                                                                                                                                                                                                                                                                                                                                                                                                          |
| *Notes*             | Includes bibliographical references and index.                                                                                                                                                                                                                                                                                                                                                                                                                              |
| *Additional formats* | Print version: Ergül, Özgür. Introduction to electromagnetic waves with Maxwell's equations Hoboken, NJ: Wiley, 2021 9781119626725 (DLC) 2021014304                                                                                                                                                                                                                                                                                                                          |

## Introduction to electromagnetic waves with Maxwell's equations

|                        |                                                                                                                                                                                                                                                                                                                                                                                                                          |
| ---------------------- | ------------------------------------------------------------------------------------------------------------------------------------------------------------------------------------------------------------------------------------------------------------------------------------------------------------------------------------------------------------------------------------------------------------------------ |
| *LCCN*                 | 2021014304                                                                                                                                                                                                                                                                                                                                                                                                               |
| *Type of material*     | Book                                                                                                                                                                                                                                                                                                                                                                                                                     |
| *Personal name*        | Ergül, Özgür, author.                                                                                                                                                                                                                                                                                                                                                                                                    |
| *Main title*           | Introduction to electromagnetic waves with Maxwell's equations / Ozgur Ergul.                                                                                                                                                                                                                                                                                                                                            |
| *Published/Produced*   | Hoboken, NJ: Wiley, 2021.                                                                                                                                                                                                                                                                                                                                                                                                |
| *ISBN*                 | 9781119626725 (cloth)                                                                                                                                                                                                                                                                                                                                                                                                    |
|                        | (adobe pdf)                                                                                                                                                                                                                                                                                                                                                                                                              |
|                        | (epub)                                                                                                                                                                                                                                                                                                                                                                                                                   |
| *LC classification*    | QC661 .E694 2021                                                                                                                                                                                                                                                                                                                                                                                                         |
| *Summary*              | "Written in an accessible style, and based on the author's extensive experience of teaching electromagnetics, this text "simply" presents the well-known topics that all electromagnetics instructors would include in their syllabus. The text is carefully tuned to be relevant to an audience of engineering students who have been exposed to the basic curriculum in linear algebra and multivariate calculus. Maxwell's equations form the backbone of the book; as they are developed step by step in |

|  |  |
|---|---|
| | consecutive chapters, related electromagnetic phenomena, as well as accompanying mathematical tools, are discussed simultaneously. Whilst the book will appeal to both undergraduate and post/graduate students, no prerequisite electromagnetics knowledge is required, as it starts with the most fundamental concepts. However, it is expected that the reader is familiar with basic calculus."-- Provided by publisher. |
| *LC Subjects* | Electromagnetic waves.<br>Maxwell equations. |
| *Notes* | Includes bibliographical references and index. |
| *Additional formats* | Online version: Ergül, Özgür, Introduction to electromagnetic waves with Maxwell's equations Hoboken, NJ: Wiley, 2021. 9781119626732 (DLC) 2021014305 |

## Let's ride a wave!: diving into the science of light and sound waves with physics

|  |  |
|---|---|
| *LCCN* | 2019059952 |
| *Type of material* | Book |
| *Personal name* | Ferrie, Chris, author. |
| *Main title* | Let's ride a wave!: diving into the science of light and sound waves with physics / Chris Ferrie. |
| *Published/Produced* | Naperville, Il.: Sourcebooks Explore, [2020] |
| *ISBN* | 9781492680581 (hardcover) |
| *LC classification* | QC174.2 .F47 2020 |
| *Summary* | "Red Kangaroo is having a fun day at the beach! She loves watching the waves go up and down, but, suddenly, she has a question: Do the waves ever stop? Dr. Chris has the answer-not only do waves never stop, but there are all kinds of waves around us! Discover all the waves you can and can't see, plus the waves you can hear! In this new series, Chris Ferrie answers all the questions Red Kangaroo has about what things are made of and how things work using real-world and practical examples. Young readers will have a firm grasp of scientific and mathematical |

|  |  |
|---|---|
| | concepts to help answer many of their "why" questions"-- Provided by publisher. |
| *LC Subjects* | Wave mechanics--Juvenile literature. |
| | Electromagnetic waves--Juvenile literature. |
| | Wave theory of light--Juvenile literature. |
| | Sound-waves--Juvenile literature. |
| *Intended Audience* | Ages 4-8. (Sourcebooks Explore.) |
| | Grades 2-3. (Sourcebooks Explore.) |

**Light-matter interaction: physics and engineering at the nanoscale**

|  |  |
|---|---|
| LCCN | 2016945264 |
| *Type of material* | Book |
| *Personal name* | Weiner, John, 1943- author. |
| *Main title* | Light-matter interaction: physics and engineering at the nanoscale / John Weiner, Frederico Nunes. |
| *Edition* | Second edition. |
| *Published/Produced* | Oxford; New York, New York: Oxford University Press, 2017. |
| | ©2017 |
| *Description* | xiv, 418 pages: illustrations, portraits; 25 cm |
| *ISBN* | 0198796668 hardback |
| | 9780198796664 hardback |
| | 0198796676 paperback |
| | 9780198796671 paperback |
| *LC classification* | TA1530 .W45 2017 |
| *Related names* | Nunes, Frederico, author. |
| *Contents* | Historical synopsis of light-matter interaction - Elements of classical electrodynamics - Physical optics of plane waves - Energy flow in polarizable matter - The classical charged oscillator and the dipole antenna - Classical black-body radiation - Surface waves - Transmission lines and waveguides - Metamaterials - Momentum in fields and matter - Atom-light forces - Radiation in classical and quantal atoms - Appendices. Appendix A. Numerical constants and dimensions - Appendix B. Systems of units in electromagnetism - Appendix C. Review of vector calculus - Appendix D. Gradient, divergence, and curl in cylindrical and polar coordinates - |

# Bibliography 143

|              |                                                                                                                                                                                                      |
| ------------ | ---------------------------------------------------------------------------------------------------------------------------------------------------------------------------------------------------- |
|              | Appendix E. Properties of phasors - Appendix F. Properties of the Laguerre functions - Appendix G. Properties of the Legendre functions - Appendix H. Properties of the Hermite polynomials. |
| *LC Subjects* | Nanophotonics. Electromagnetic theory. Electromagnetic surface waves. Plasmons (Physics) Electric circuits. |
| *Other Subjects* | Electric circuits. Electromagnetic surface waves. Electromagnetic theory. Nanophotonics. Plasmons (Physics) |
| *Notes* | Includes bibliographical references and index. |

## Linear and nonlinear wave propagation

| | |
|---|---|
| *LCCN* | 2021427279 |
| *Type of material* | Book |
| *Personal name* | Kuo, Spencer P., author. |
| *Main title* | Linear and nonlinear wave propagation / Spencer Kuo, New York University, USA. |
| *Published/Produced* | New Jersey: World Scientific, [2021] |
| *Description* | xvii, 187 pages: illustrations; 24 cm |
| *ISBN* | 9789811231636 hardcover |
| | 981123163X hardcover |
| *LC classification* | QA927 .K86 2021 |
| *LC Subjects* | Wave-motion, Theory of. Electromagnetic waves. |
| *Other Subjects* | Electromagnetic waves. Wave-motion, Theory of. |
| *Notes* | Includes bibliographical references and index. |

## Modern characterization of electromagnetic systems and its associated metrology

| | |
|---|---|
| *LCCN* | 2020008265 |
| *Type of material* | Book |
| *Main title* | Modern characterization of electromagnetic systems and its associated metrology / edited by Tapan K. |

|   |   |
|---|---|
|  | Sarkar, Magdalena Salazar-Palma, Ming Da Zhu, Heng Chen. |
| *Published/Produced* | Hoboken, NJ: Wiley, 2020. |
| *Description* | 1 online resource |
| *ISBN* | 9781119076537 (epub) |
|  | 9781119076544 (adobe pdf) |
|  | (hardback) |
| *LC classification* | QC760 |
| *Related names* | Sarkar, Tapan (Tapan K.), editor. |
|  | Salazar-Palma, Magdalena, editor. |
|  | Zhu, Ming Da, editor. |
|  | Chen, Heng, 1990- editor. |
| *Summary* | "This book describes new method of characterization of electromagnetic wave dynamics and measurement, which are based on modern computational and digital signal processing techniques. The book introduces modern computational concepts in electromagnetic system characterization and introduce modern signal processing algorithms not only to enhance the resolution but also extract information from electromagnetic systems that is not currently possible, for example, generation of the non-minimum phase or for that matter the transient response given amplitude only data. The author covers model based parameter estimation and planar near field to far field transformation, as well as spherical near field to far field transformation. Electromagnetism is the physics of the electromagnetic field: a field, encompassing all of space, which exerts a force on those particles that possess a property known as electric charge, and is in turn affected by the presence and motion of such particles. The design of circuits that use the electromagnetic properties of electrical components such as resistors, capacitors, inductors, diodes and transistors to achieve a particular functionality."-- Provided by publisher. |
| *Contents* | Mathamatical Principles related to Modern system Analysis - Matrix Pencil Method (MPM) - The |

|  |  |
|---|---|
|  | Cauchy Method - Applications of the Hilbert Transform: A Nonparametric method for Interpolation/Extrapolation of data - The Source Reconstruction Method - Planar near-field to far-field transformation using a single moving probe and a fixed probe arrays - Spherical Near-field to Far-field transformation - Deconvolving Measured Electromagnetic Responses - Performance of Different Functionals for Interpolation /Extrapolation of Near/Far Field Data - Retrieval of free space radiation patterns from measured data in a non-anechoic environment. |
| *LC Subjects* | Electromagnetism--Mathematics. |
|  | Electromagnetic waves--Measurement. |
| *Notes* | Includes bibliographical references and index. |
| *Additional formats* | Print version: Modern characterization of electromagnetic systems and its associated metrology Hoboken, NJ: Wiley, 2020. 9781119076469 (DLC) 2020008264 |

**Modern characterization of electromagnetic systems and its associated metrology**

|  |  |
|---|---|
| *LCCN* | 2020008264 |
| *Type of material* | Book |
| *Main title* | Modern characterization of electromagnetic systems and its associated metrology / edited by Tapan K. Sarkar, Magdalena Salazar-Palma, Ming Da Zhu, Heng Chen. |
| *Published/Produced* | Hoboken, NJ: Wiley, 2020. |
| *ISBN* | 9781119076469 (hardback) |
|  | (adobe pdf) |
|  | (epub) |
| *LC classification* | QC760 .M53 2020 |
| *Related names* | Sarkar, Tapan (Tapan K.), editor. |
|  | Salazar-Palma, Magdalena, editor. |
|  | Zhu, Ming Da, editor. |
|  | Chen, Heng, 1990- editor. |
| *Summary* | "This book describes new method of characterization of electromagnetic wave dynamics and measurement, |

which are based on modern computational and digital signal processing techniques. The book introduces modern computational concepts in electromagnetic system characterization and introduce modern signal processing algorithms not only to enhance the resolution but also extract information from electromagnetic systems that is not currently possible, for example, generation of the non-minimum phase or for that matter the transient response given amplitude only data. The author covers model based parameter estimation and planar near field to far field transformation, as well as spherical near field to far field transformation. Electromagnetism is the physics of the electromagnetic field: a field, encompassing all of space, which exerts a force on those particles that possess a property known as electric charge, and is in turn affected by the presence and motion of such particles. The design of circuits that use the electromagnetic properties of electrical components such as resistors, capacitors, inductors, diodes and transistors to achieve a particular functionality."--Provided by publisher.

*Contents*  Mathamatical Principles related to Modern system Analysis - Matrix Pencil Method (MPM) - The Cauchy Method - Applications of the Hilbert Transform: A Nonparametric method for Interpolation/Extrapolation of data - The Source Reconstruction Method - Planar near-field to far-field transformation using a single moving probe and a fixed probe arrays - Spherical Near-field to Far-field transformation - Deconvolving Measured Electromagnetic Responses - Performance of Different Functionals for Interpolation /Extrapolation of Near/Far Field Data - Retrieval of free space radiation patterns from measured data in a non-anechoic environment.

*LC Subjects*  Electromagnetism--Mathematics.
Electromagnetic waves--Measurement.

| | |
|---|---|
| *Notes* | Includes bibliographical references and index. |
| *Additional formats* | Online version: Modern characterization of electromagnetic systems and its associated metrology Hoboken, NJ: Wiley, 2020. 9781119076544 (DLC) 2020008265 |

**Modern electromagnetic scattering theory with applications**

| | |
|---|---|
| *LCCN* | 2016028695 |
| *Type of material* | Book |
| *Personal name* | Osipov, Andrey (Andrey V.) |
| *Main title* | Modern electromagnetic scattering theory with applications / Andrey V. Osipov, Microwaves and Radar Institute, German Aerospace Center (DLR), Germany, Sergei A. Tretyakov, School of Electrical Engineering, Aalto University, Finland. |
| *Published/Produced* | Chichester, West Sussex: Wiley, 2017. |
| *Description* | xviii, 806 pages: illustrations; 26 cm |
| *ISBN* | 9780470512388 (cloth) |
| *LC classification* | QC665.S3 O85 2017 |
| *Related names* | Tretyakov, Sergei. |
| *LC Subjects* | Electromagnetic waves--Scattering. Electromagnetic fields. Radar cross sections. |
| *Notes* | Includes bibliographical references and index. |

**Optical effects in solids**

| | |
|---|---|
| *LCCN* | 2018046542 |
| *Type of material* | Book |
| *Personal name* | Tanner, David B., 1945- author. |
| *Main title* | Optical effects in solids / David B. Tanner, University of Florida. |
| *Published/Produced* | Cambridge; New York, NY: Cambridge University Press, [2019] |
| *Description* | xii, 400 pages: illustrations (some color); 26 cm |
| *ISBN* | 9781107160149 (hardback) |
| *LC classification* | QC176.8.O6 T36 2019 |
| *LC Subjects* | Solids--Optical properties. Light. Materials--Optical properties. |

|   |   |
|---|---|
|  | Light--Scattering. |
|  | Reflection (Optics) |
|  | Electromagnetic waves. |
| Notes | Includes bibliographical references and index. |

### Polarized light and the Mueller matrix approach

|   |   |
|---|---|
| LCCN | 2021050506 |
| Type of material | Book |
| Personal name | Gil Pérez, José Jorge, author. |
| Main title | Polarized light and the Mueller matrix approach / José J. Gil, Razvigor Ossikovski. |
| Edition | Second edition. |
| Published/Produced | Boca Raton: CRC Press, 2022. |
| ISBN | 9780367407469 (hardback) |
|  | 9781032215112 (paperback) |
|  | (ebook) |
| LC classification | QC441 .G55 2022 |
| Related names | Ossikovski, Razvigor, author. |
| Summary | "An Up-to-Date Compendium on the Physics and Mathematics of Polarization Phenomena Now thoroughly revised, Polarized Light and the Mueller Matrix Approach cohesively integrates basic concepts of polarization phenomena from the dual viewpoints of the states of polarization of electromagnetic waves and the transformations of these states by the action of material media. Through selected examples, it also illustrates actual and potential applications in materials science, biology, and optics technology. The book begins with the basic concepts related to two- and three-dimensional polarization states. It next describes the nondepolarizing linear transformations of the states of polarization through the Jones and Mueller-Jones approaches. The authors then discuss the forms and properties of the Jones and Mueller matrices associated with different types of nondepolarizing media, address the foundations of the Mueller matrix, and delve more deeply into the analysis of the physical parameters associated with Mueller |

matrices. The authors proceed with introducing the arbitrary decomposition and other useful parallel decompositions, and compare the powerful serial decompositions of depolarizing Mueller matrices. They also analyze the general formalism and specific algebraic quantities and notions related to the concept of differential Mueller matrix. Useful approaches that provide a geometric point of view on the polarization effects exhibited by different types of media are also comprehensively described. The book concludes with a new chapter devoted to the main procedures for filtering measured Mueller matrices. Suitable for advanced graduates and more seasoned professionals, this book covers the main aspects of polarized radiation and polarization effects of material media. It expertly combines physical and mathematical concepts with important approaches for representing media through equivalent systems composed of simple components"-- Provided by publisher.

| | |
|---|---|
| *LC Subjects* | Electromagnetic waves--Polarization. |
| | Polarization (Light) |
| *Notes* | Includes bibliographical references and index. |
| *Additional formats* | Online version: Gil, José J., 1956- Polarized light and the mueller matrix approach 2. Boca Raton: CRC Press, 2022 9780367815578 (DLC) 2021050507 |
| *Series* | Series in optics and optoelectronics |

**Polarized light and the Mueller matrix approach**

| | |
|---|---|
| *LCCN* | 2021050507 |
| *Type of material* | Book |
| *Personal name* | Gil Pérez, José Jorge, author. |
| *Main title* | Polarized light and the Mueller matrix approach / José J. Gil, Razvigor Ossikovski. |
| *Edition* | Second edition. |
| *Published/Produced* | Boca Raton: CRC Press, 2022. |
| *Description* | 1 online resource |
| *ISBN* | 9780367815578 (ebook) |
| | (hardback) |
| | (paperback) |

| | |
|---|---|
| *LC classification* | QC441 |
| *Related names* | Ossikovski, Razvigor, author. |
| *Summary* | "An Up-to-Date Compendium on the Physics and Mathematics of Polarization Phenomena Now thoroughly revised, Polarized Light and the Mueller Matrix Approach cohesively integrates basic concepts of polarization phenomena from the dual viewpoints of the states of polarization of electromagnetic waves and the transformations of these states by the action of material media. Through selected examples, it also illustrates actual and potential applications in materials science, biology, and optics technology. The book begins with the basic concepts related to two- and three-dimensional polarization states. It next describes the nondepolarizing linear transformations of the states of polarization through the Jones and Mueller-Jones approaches. The authors then discuss the forms and properties of the Jones and Mueller matrices associated with different types of nondepolarizing media, address the foundations of the Mueller matrix, and delve more deeply into the analysis of the physical parameters associated with Mueller matrices. The authors proceed with introducing the arbitrary decomposition and other useful parallel decompositions, and compare the powerful serial decompositions of depolarizing Mueller matrices. They also analyze the general formalism and specific algebraic quantities and notions related to the concept of differential Mueller matrix. Useful approaches that provide a geometric point of view on the polarization effects exhibited by different types of media are also comprehensively described. The book concludes with a new chapter devoted to the main procedures for filtering measured Mueller matrices. Suitable for advanced graduates and more seasoned professionals, this book covers the main aspects of polarized radiation and polarization effects of material media. It expertly combines physical and mathematical |

Bibliography

|  |  |
|---|---|
|  | concepts with important approaches for representing media through equivalent systems composed of simple components"-- Provided by publisher. |
| LC Subjects | Electromagnetic waves--Polarization. |
|  | Polarization (Light) |
| Notes | Includes bibliographical references and index. |
| Additional formats | Print version: Gil Pérez, José Jorge. Polarized light and the Mueller matrix approach Second edition. Boca Raton: CRC Press, 2022 9780367407469 (DLC) 2021050506 |
| Series | Series in optics and optoelectronics |

**Principles of electromagnetic waves and materials**

| LCCN | 2017015930 |
|---|---|
| Type of material | Book |
| Personal name | Kalluri, Dikshitulu K., author. |
| Main title | Principles of electromagnetic waves and materials / Dikshitulu K. Kalluri. |
| Edition | Second edition. |
| Published/Produced | Boca Raton: Taylor & Francis, CRC Press, 2018. |
| ISBN | 9781498733298 (hardback) |
| LC classification | QC760 .K363 2018 |
| LC Subjects | Electromagnetism--Mathematical models. Electromagnetic waves--Computer simulation. |
| Notes | Includes bibliographical references and index. |

**Radio waves as invisible gift of our awesome God: explorations in the science of astronomy**

| LCCN | 2020392231 |
|---|---|
| Type of material | Book |
| Personal name | Ogwo, Jemima Ngozi, author. |
| Main title | Radio waves as invisible gift of our awesome God: explorations in the science of astronomy / by Jemima Ngozi Ogwo. |
| Published/Produced | Uturu, [Nigeria]: Abia State University, [2018] |
| Description | xvi, 50 pages: illustrations (some color); 21 cm |
| LC classification | QB478.5 .O398 2018 |
| LC Subjects | Radio astronomy. Radio waves. |

|             |                                                                 |
|-------------|-----------------------------------------------------------------|
| *Notes*     | Electromagnetic theory. "Delivered on 18th July 2018".         |
| *Series*    | Inaugural lecture of Abia State University, Uturu, Abia State, Nigeria; 40th |

**Resistivity modeling: propagation, laterolog and micro-pad analysis**

| | |
|---|---|
| *LCCN* | 2016042818 |
| *Type of material* | Book |
| *Personal name* | Chin, Wilson C., author. |
| *Main title* | Resistivity modeling: propagation, laterolog and micro-pad analysis / Wilson C. Chin, Ph.D., M.I.T. |
| *Published/Produced* | Hoboken, New Jersey: John Wiley & Sons, Inc.,; Beverly, Massachusetts: Scrivener Publishing LLC, [2017] |
| *Description* | xvii, 295 pages: illustrations; 24 cm |
| *ISBN* | 9781118925997 (hardback) |
| *LC classification* | TN871.35 .C477 2017 |
| *Contents* | Physics, math and basic ideas - Axisymmetric transient models - Steady axisymmetric formulations - Direct current models for micro-pad devices - Coil antenna modeling for MWD applications - What is resistivity? - Multiphase flow and transient resistivity - Analytical methods for time lapse well logging analysis. |
| *LC Subjects* | Oil well logging, Electric--Mathematical models. Electromagnetic waves--Mathematical models. |
| *Notes* | Includes bibliographical references (pages 272-275) and index. |

**Scooby-Doo!, a science of light mystery: the angry alien**

| | |
|---|---|
| *LCCN* | 2016054932 |
| *Type of material* | Book |
| *Personal name* | Peterson, Megan Cooley, author. |
| *Main title* | Scooby-Doo!, a science of light mystery: the angry alien / by Megan Cooley Peterson; illustrated by Dario Brizuela. |
| *Published/Produced* | North Mankato, Minnesota: Capstone Press, a Capstone imprint, 2017. |
| *Description* | 32 pages: color illustrations; 24 cm. |

Bibliography 153

| | |
|---|---|
| *ISBN* | 9781515737001 (library binding) |
| | 9781515737049 (pbk.) |
| *LC classification* | QC360 .P435 2017 |
| *Portion of title* | Science of light mystery |
| | Angry alien |
| *Related names* | Brizuela, Dario, illustrator. |
| *LC Subjects* | Scooby-Doo (Fictitious character)--Juvenile literature. |
| | Light--Juvenile literature. |
| | Electromagnetic waves--Juvenile literature. |
| *Notes* | Includes bibliographical references and index. |
| *Intended Audience* | Ages 9-12. |
| | Grades 4 to 6. |
| *Series* | Scooby-Doo!: a science of light mystery |

**The science of light waves**

| | |
|---|---|
| *LCCN* | 2016059051 |
| *Type of material* | Book |
| *Personal name* | Johnson, Robin (Robin R.), author. |
| *Main title* | The science of light waves / Robin Johnson. |
| *Published/Produced* | St. Catharines, Ontario; New York, New York: Crabtree Publishing Company, [2017] |
| *Description* | 32 pages: color illustrations; 28 cm. |
| *ISBN* | 9780778729440 (reinforced library binding) |
| | 9780778729686 (pbk.) |
| *LC classification* | QC403 .J64 2017 |
| *Summary* | "This engaging book describes the properties of light waves, how they move, and the way our eyes receive them. Readers will learn that we see an object when light reflects from its surface and into the eye. A link to interactive activities online plus an activity in the book allow readers to create models that explore how to redirect and block the path a light wave travels."-- Provided by publisher. |
| *LC Subjects* | Wave theory of light--Juvenile literature. |
| | Electromagnetic waves--Juvenile literature. |
| | Light--Juvenile literature. |
| *Notes* | Includes bibliographical references and index. |
| *Intended Audience* | Ages 8-11. |

| | |
|---|---|
| *Additional formats* | Grades 4 to 6.<br>Online version: Johnson, Robin (Robin R.), author. Science of light waves New York, New York: Crabtree Publishing Company, [2017] 9781427118561 (DLC) 2016059909 |

**Tuning into frequency: the invisible force that heals us and the planet**

| | |
|---|---|
| *LCCN* | 2020021358 |
| *Type of material* | Book |
| *Corporate name* | Sputnik Futures, author. |
| *Main title* | Tuning into frequency: the invisible force that heals us and the planet / Sputnik Futures. |
| *Edition* | First Tiller Press trade paperback edition. |
| *Published/Produced* | New York: Tiller Press, 2021. |
| *Description* | 1 online resource |
| *ISBN* | 9781982147952 (ebook)<br>(paperback) |
| *LC classification* | RZ421 |
| *Summary* | "A riveting guide to the energy that surrounds us-from the vitality of light to the potency of sound waves-and how tuning into the power of frequencies can help us heal ourselves, and the planet"-- Provided by publisher. |
| *Contents* | Introduction: Good Vibrations - We Live In An Electromagnetic World And I Am an Electromagnetic Girl - Catching Nature's Vibe: The Wild, Wonderful World of Nature's Internet - The Body Light: The Healing Spectrum From The Cosmos to Color - Frequency Healing: Turn On, Tune In, And Take Your Electromedicine - Free Energy and The Quest for The Zero Point - The Mind Field: PSI Science and The Flow of Information - Alice In Futureland: A Speculative Life In 2050 - Down The Rabbit Hole: Articles, Further Reading, People, Organization. |
| *LC Subjects* | Energy medicine.<br>Light--Therapeutic use.<br>Vibration--Therapeutic use.<br>Sound-waves--Therapeutic use. |

|  |  |
|---|---|
| Notes | Electromagnetic fields--Therapeutic use.<br>Includes bibliographical references and index. |
| Additional formats | Print version: Sputnik Futures. Tuning into frequency First Tiller Press trade paperback edition. New York: Tiller Press, 2021. 9781982147945 (DLC) 2020021357 |

**Tuning into frequency: the invisible force that heals us and the planet**

|  |  |
|---|---|
| LCCN | 2020021357 |
| Type of material | Book |
| Corporate name | Sputnik Futures, author. |
| Main title | Tuning into frequency: the invisible force that heals us and the planet / Sputnik Futures. |
| Edition | First Tiller Press trade paperback edition. |
| Published/Produced | New York: Tiller Press, 2021. |
| ISBN | 9781982147945 (paperback)<br>(ebook) |
| LC classification | RZ421 .S68 2021 |
| Summary | "A riveting guide to the energy that surrounds us-from the vitality of light to the potency of sound waves-and how tuning into the power of frequencies can help us heal ourselves, and the planet"-- Provided by publisher. |
| Contents | Introduction: Good Vibrations - We Live In An Electromagnetic World And I Am an Electromagnetic Girl - Catching Nature's Vibe: The Wild, Wonderful World of Nature's Internet - The Body Light: The Healing Spectrum From The Cosmos to Color - Frequency Healing: Turn On, Tune In, And Take Your Electromedicine - Free Energy and The Quest for The Zero Point - The Mind Field: PSI Science and The Flow of Information - Alice In Futureland: A Speculative Life In 2050 - Down The Rabbit Hole: Articles, Further Reading, People, Organization. |
| LC Subjects | Energy medicine.<br>Light--Therapeutic use.<br>Vibration--Therapeutic use.<br>Sound-waves--Therapeutic use. |

|  |  |
|---|---|
| | Electromagnetic fields--Therapeutic use. |
| *Notes* | Includes bibliographical references and index. |
| *Additional formats* | Online version: Sputnik Futures Tuning into frequency First Tiller Press trade paperback edition. New York: Tiller Press, 2020. 9781982147952 (DLC) 2020021358 |
| *Series* | Alice in futureland |

## Understanding and using X-rays

|  |  |
|---|---|
| *LCCN* | 2019015484 |
| *Type of material* | Book |
| *Personal name* | Rubio, Elizabeth, author. |
| *Main title* | Understanding and using X-rays / Elizabeth Rubio. |
| *Published/Produced* | New York, NY: Enslow Publishing, 2020. ©2021 |
| *ISBN* | 9781978515017 (library bound) 9781978515000 (pbk.) |
| *LC classification* | QC481 .R83 2020 |
| *Summary* | "This book is about x-rays."-- Provided by publisher. |
| *Contents* | The electromagnetic spectrum - The history of X-rays - Medical uses of X-rays - Harm from X-rays - Nuclear weaponry - X-rays and scientific discovery - Other uses of X-rays. |
| *LC Subjects* | X-rays--Juvenile literature. Electromagnetic waves--Juvenile literature. |
| *Notes* | Includes bibliographical references and index. |
| *Intended Audience* | Grades 7 to 12. |
| *Series* | The electromagnetic spectrum |

## Wave propagation in materials for modern applications

|  |  |
|---|---|
| *LCCN* | 2016946464 |
| *Type of material* | Book |
| *Main title* | Wave propagation in materials for modern applications / editor, Carlos Granger. |
| *Published/Produced* | Valley Cottage, NY: Scitus Academics LLC, [2017] |
| *Description* | viii, 282 pages: illustrations; 24 cm |
| *ISBN* | 9781681176093 |
| *LC classification* | QC662 .W38 2017 |
| *Related names* | Granger, Carlos, editor. |

| | |
|---|---|
| *LC Subjects* | Electromagnetic waves. |
| *Notes* | Includes bibliographical references and index. |

### Wave theory of information
| | |
|---|---|
| *LCCN* | 2017032961 |
| *Type of material* | Book |
| *Personal name* | Franceschetti, Massimo, author. |
| *Main title* | Wave theory of information / Massimo Franceschetti, University of California, San Diego. |
| *Published/Produced* | Cambridge: Cambridge University Press, 2017. |
| *ISBN* | 9781107022317 (hardback) |
| *LC classification* | Q360 .F73 2017 |
| *LC Subjects* | Information theory. |
| | Electromagnetic waves. |
| | Wave-motion, Theory of. |
| *Notes* | Includes bibliographical references. |

### Waves and information transfer
| | |
|---|---|
| *LCCN* | 2016059052 |
| *Type of material* | Book |
| *Personal name* | Hudak, Heather C., 1975- author. |
| *Main title* | Waves and information transfer / Heather C. Hudak. |
| *Published/Produced* | St. Catharines, Ontario; New York, New York: Crabtree Publishing Company, [2017] |
| *Description* | 32 pages: color illustrations; 28 cm. |
| *ISBN* | 9780778729624 (reinforced library binding) |
| | 9780778729709 (pbk.) |
| *LC classification* | QC661 .H785 2017 |
| *Summary* | "In this fascinating title, readers explore how light and sound waves transfer information. From telescopes that extend our sense of sight to satellites that help us communicate across large distances, patterns of waves transfer information in many ways. A link to interactive activities online plus an activity in the book allow readers to use what they have learned about waves to engineer wave patterns that communicate across distances."-- Provided by publisher. |
| *LC Subjects* | Electromagnetic waves--Juvenile literature. |

158    Bibliography

|   |   |
|---|---|
|   | Wave mechanics--Juvenile literature. |
|   | Information technology--Juvenile literature. |
| *Notes* | Includes bibliographical references and index. |
| *Intended Audience* | Ages 8-11. |
|   | Grades 4 to 6. |
| *Additional formats* | Online version: Hudak, Heather C., 1975- author. Waves and information transfer New York, New York: Crabtree Publishing Company, [2017] 9781427118578 (DLC) 2016059910 |
| *Series* | Catch a wave |
|   | Crabtree plus |

## Waves in complex media: fundamentals and device applications

| | |
|---|---|
| *LCCN* | 2021026904 |
| *Type of material* | Book |
| *Personal name* | Dal Negro, Luca, author. |
| *Main title* | Waves in complex media: fundamentals and device applications / Luca Dal Negro. |
| *Published/Produced* | New York: Cambridge university Press, 2021. |
| *Description* | 1 online resource |
| *ISBN* | 9781139775328 (epub) |
|   | (hardback) |
| *LC classification* | QC670 |
| *Summary* | "Why should we learn about the behavior of waves in optical media with irregular, non-periodic, and even disordered structures? First of all, because they can be found everywhere around us, from complex functional materials to the arrangement of leaves on plant stems and even at the inner core of number theory! Second, because the scattering behavior of waves in complex media surprises us with emergent phenomena driven by interference effects. Third, because waves in complex media unveil profound analogies between the classical and quantum transport regimes beyond standard diffusion theory such as, for example, Anderson light localization. However, while waves in periodic structures have been deeply investigated for more than a century and are discussed by the majority of graduate-level |

|  |  |
|---|---|
|  | textbooks in optics and photonics, the behavior of waves in more general aperiodic environments is almost exclusively addressed in the specialized literature and consequently has limited impact on graduate curricula. It is my goal to bridge this gap by offering a comprehensive and interdisciplinary textbook that systematically addresses both the conceptual foundation and the scattering properties of optical waves in complex media"-- Provided by publisher. |
| *LC Subjects* | Electromagnetic waves. |
|  | Electromagnetic waves--Scattering. |
| *Other Subjects* | Technology & Engineering / Electronics / Optoelectronics |
| *Notes* | Includes bibliographical references and index. |
| *Additional formats* | Print version: Dal Negro, Luca. Waves in complex media New York: Cambridge university Press, 2021 9781107037502 (DLC) 2021026903 |

**Waves in complex media: fundamentals and device applications**

|  |  |
|---|---|
| *LCCN* | 2021026903 |
| *Type of material* | Book |
| *Personal name* | Dal Negro, Luca, author. |
| *Main title* | Waves in complex media: fundamentals and device applications / Luca Dal Negro. |
| *Published/Produced* | New York: Cambridge university Press, 2021. |
| *ISBN* | 9781107037502 (hardback) |
|  | (epub) |
| *LC classification* | QC670 .D34 2021 |
| *Summary* | "Why should we learn about the behavior of waves in optical media with irregular, non-periodic, and even disordered structures? First of all, because they can be found everywhere around us, from complex functional materials to the arrangement of leaves on plant stems and even at the inner core of number theory! Second, because the scattering behavior of waves in complex media surprises us with emergent phenomena driven by interference effects. Third, because waves in complex media unveil profound |

|  |  |
|---|---|
|  | analogies between the classical and quantum transport regimes beyond standard diffusion theory such as, for example, Anderson light localization. However, while waves in periodic structures have been deeply investigated for more than a century and are discussed by the majority of graduate-level textbooks in optics and photonics, the behavior of waves in more general aperiodic environments is almost exclusively addressed in the specialized literature and consequently has limited impact on graduate curricula. It is my goal to bridge this gap by offering a comprehensive and interdisciplinary textbook that systematically addresses both the conceptual foundation and the scattering properties of optical waves in complex media"-- Provided by publisher. |
| *LC Subjects* | Electromagnetic waves. |
|  | Electromagnetic waves--Scattering. |
| *Other Subjects* | Technology & Engineering / Electronics / Optoelectronics |
| *Notes* | Includes bibliographical references and index. |
| *Additional formats* | Online version: Dal Negro, Luca, 1975- Waves in complex media 1. New York: Cambridge university Press, 2021 9781139775328 (DLC) 2021026904 |

**Weather radar polarimetry**

| | |
|---|---|
| *LCCN* | 2016005615 |
| *Type of material* | Book |
| *Personal name* | Zhang, Guifu, 1962- |
| *Main title* | Weather radar polarimetry / Guifu Zhang. |
| *Published/Produced* | Boca Raton: CRC Press, Taylor & Francis Group, [2017] |
| *Description* | xvii, 304 pages: illustrations (some color), maps (chiefly color); 25 cm |
| *ISBN* | 9781439869581 (hbk.: acid-free paper) |
| *LC classification* | QC973.5 .Z43 2017 |
| *LC Subjects* | Radar meteorology. |
|  | Polarimetry. |
|  | Radar--Interference. |

# Bibliography

|  |  |
|---|---|
| Notes | Electromagnetic waves--Scattering. Remote sensing. Includes bibliographical references (pages 283-296) and index. |

### What are waves?

|  |  |
|---|---|
| LCCN | 2016059053 |
| Type of material | Book |
| Personal name | Hudak, Heather C., 1975- author. |
| Main title | What are waves? / Heather C. Hudak. |
| Published/Produced | St. Catharines, Ontario; New York, New York: Crabtree Publishing Company, [2017] |
| Description | 32 pages: color illustrations; 28 cm. |
| ISBN | 9780778729648 (reinforced library binding) 9780778729723 (pbk.) |
| LC classification | QC174.2 .H83 2017 |
| Summary | "This exciting title introduces readers to the concept of a wave and the patterns and properties common to both light and sound waves. Clear text and detailed diagrams combine to demonstrate the cause-and-effect relationships involved in the properties of amplitude, wavelength, and frequency. A link to interactive activities online plus an activity in the book allow readers to explore key concepts close up by creating their own wave models."-- Provided by publisher. |
| LC Subjects | Wave mechanics--Juvenile literature. Electromagnetic waves--Juvenile literature. Wave-motion, Theory of--Juvenile literature. |
| Notes | Includes bibliographical references and index. |
| Intended Audience | Ages 8-11. Grades 4 to 6. |
| Additional formats | Online version: Hudak, Heather C., 1975- author. What are waves? New York, New York: Crabtree Publishing Company, [2017] 9781427118585 (DLC) 2016059911 |
| Series | Catch a wave Crabtree plus |

**What the ear hears (and doesn't): inside the extraordinary everyday world of frequency**

| | |
|---|---|
| LCCN | 2022026567 |
| Type of material | Book |
| Personal name | Mainwaring, Richard, author. |
| Main title | What the ear hears (and doesn't): inside the extraordinary everyday world of frequency / Richard Mainwaring. |
| Published/Produced | Naperville, Illinois: Sourcebooks, [2022] |
| Description | 1 online resource |
| ISBN | 9781728259376 (adobe pdf) |
| | 9781728259383 (epub) |
| | (trade paperback) |
| LC classification | QC229 |
| Summary | "In 2011, without warning, a skyscraper in South Korea began to shake uncontrollably and was immediately evacuated. Was it an earthquake? A terrorist attack? No one seemed quite sure. The actual cause emerged later: Twenty-three middle-aged Koreans were having a Tae Bo fitness class in the office gym on the twelfth floor. Their beats had inadvertently matched the building's natural frequency, and this coincidence caused the building to shake at an alarming rate for ten minutes. Frequency is all around us, but really isn't understood. What the Ear Hears (and Doesn't) reveals the extraordinary world of frequency-from medicine to religion to the environment to the paranormal-not through abstract theory, but through a selection of small memorable human (and animal) stories laced with dry humor, including: The elephant who anticipated the 2004 Asian tsunami and carried a young girl miles inland, saving her life The reason for those deep spiritual feelings we have in churches The cutting-edge methods that are changing medicine The world's loneliest whale"-- Provided by publisher. |
| LC Subjects | Sound-waves--Miscellanea |
| | Sound--Miscellanea |
| | Electromagnetic waves--Miscellanea, |

| | |
|---|---|
| *Notes* | Includes bibliographical references and index. |
| *Additional formats* | Print version: Mainwaring, Richard. What the ear hears (and doesn't) Naperville, Illinois: Sourcebooks, [2022] 9781728259369 (DLC) 2022026566 |

## What the ear hears (and doesn't): inside the extraordinary everyday world of frequency

| | |
|---|---|
| *LCCN* | 2022026566 |
| *Type of material* | Book |
| *Personal name* | Mainwaring, Richard, author. |
| *Main title* | What the ear hears (and doesn't): inside the extraordinary everyday world of frequency / Richard Mainwaring. |
| *Published/Produced* | Naperville, Illinois: Sourcebooks, [2022] |
| *ISBN* | 9781728259369 (trade paperback) |
| | (epub) |
| | (adobe pdf) |
| *LC classification* | QC229 .M35 2022 |
| *Summary* | "In 2011, without warning, a skyscraper in South Korea began to shake uncontrollably and was immediately evacuated. Was it an earthquake? A terrorist attack? No one seemed quite sure. The actual cause emerged later: Twenty-three middle-aged Koreans were having a Tae Bo fitness class in the office gym on the twelfth floor. Their beats had inadvertently matched the building's natural frequency, and this coincidence caused the building to shake at an alarming rate for ten minutes. Frequency is all around us, but really isn't understood. What the Ear Hears (and Doesn't) reveals the extraordinary world of frequency-from medicine to religion to the environment to the paranormal-not through abstract theory, but through a selection of small memorable human (and animal) stories laced with dry humor, including: The elephant who anticipated the 2004 Asian tsunami and carried a young girl miles inland, saving her life The reason for those deep spiritual feelings we have in churches The |

|  |  |
|---|---|
|  | cutting-edge methods that are changing medicine The world's loneliest whale"-- Provided by publisher. |
| LC Subjects | Sound-waves--Miscellanea |
|  | Sound--Miscellanea |
|  | Electromagnetic waves--Miscellanea, |
| Notes | Includes bibliographical references and index. |
| Additional formats | Online version: Mainwaring, Richard. What the ear hears (and doesn't) Naperville, Illinois: Sourcebooks, [2022] 9781728259383 (DLC) 2022026567 |

**Workshop on Astrophysical Opacities: proceedings of a workshop held at Fetzer Center, Western Michigan University, Kalamazoo, MI, 1-4 August 2017**

|  |  |
|---|---|
| LCCN | 2018942467 |
| Type of material | Book |
| Meeting name | Workshop on Astrophysical Opacities (2017: Kalamazoo, Mich.), author. |
| Main title | Workshop on Astrophysical Opacities: proceedings of a workshop held at Fetzer Center, Western Michigan University, Kalamazoo, MI, 1-4 August 2017 / edited by Claudio Mendoza, Sylvaine Turck-Chièze, James Colgan. |
| Edition | First edition. |
| Published/Produced | San Francisco, California: Astronomical Society of the Pacific, [2018] |
|  | ©2018 |
| Description | xv, 324 pages: illustrations; 24 cm. |
| ISBN | 9781583819142 hardbound |
|  | 1583819142 hardbound |
|  | electronic |
|  | electronic |
| LC classification | QB43.3 .W67 2017 |
| Related names | Mendoza, Claudio (Mendoza Guardia), editor. |
|  | Turck-Chièze, S., editor. |
|  | Colgan, James (James Patrick), editor. |
| Contents | Stellar optics and the solar abundance problem - Computation and measurement of atomic opacities - Computation and measurement of molecular opacities - Stellar atomic diffusion - Stellar models - |

|   |   |
|---|---|
| LC Subjects | Brown dwarfs, exoplanets, and protoplanetary disks - Astrophysical processes and atomic data - Epilogue. Opacity (Optics)--Congresses. Electromagnetic waves--Diffraction--Congresses. Extrasolar planets--Observations--Congresses. Sun--Observations--Congresses. |
| Other Subjects | Electromagnetic waves--Diffraction. Observation (Scientific method) Opacity (Optics) Sun. |
| Form/Genre | Conference papers and proceedings. |
| Notes | Includes bibliographical references and index. |
| Additional formats | Online version: Workshop on Astrophysical Opacities (2017: Kalamazoo, Mich.). Workshop on Astrophysical Opacities. First edition. San Francisco, Claifornia: Astronomical Society of the Pacific, [2018] 9781583819159 |
| Series | Astronomical Society of the Pacific conference series; volume 515 Astronomical Society of the Pacific conference series; v. 515. |

# Index

## #

4G antenna, 55

## A

amplitude, 34

## B

bandgap, 33, 35, 36, 38, 52
bandwidth, viii, 2, 3, 15, 56, 64
beams, viii, 29, 46, 47, 48, 49, 52

## C

carbon, viii, 2, 3, 7, 8, 9, 11, 13, 14, 15, 16, 17, 21
cellular communication, 55
classical electrodynamics, 118, 142
cobalt spinel ferrite, 2
$CoFe_2O_4$, v, vii, 1, 2
collimation effect, 30, 44, 47, 53
composite(s), vii, 2, 3, 4, 5, 6, 7, 8, 9, 10, 11, 12, 13, 14, 15, 16, 17, 18, 19, 20, 21
composites, vii, 2, 3, 5, 6, 8, 9, 10, 11, 12, 13, 14, 16, 20, 21
crystal structure, 53
crystals, 30, 52, 53

## D

dielectric constant, 75
differential equations, 84
diffraction, ix, 67, 68, 69, 83, 89, 92, 95, 96, 97, 101, 116, 117, 118, 119, 165
directional antennas, 56

## E

electromagnetic interference shielding effectiveness, 2
electromagnetic properties, 2, 144, 146
electromagnetic wave, iii, v, vii, ix, 30, 33, 34, 52, 67, 68, 69, 71, 73, 75, 77, 79, 81, 83, 85, 86, 87, 89, 90, 91, 93, 95, 96, 97, 99, 100, 102, 106, 107, 117, 118, 120, 125, 127, 128, 132, 139, 140, 141, 144, 145, 148, 150, 151
electromagnetic waves, vii, 34, 52, 68, 69

## F

fabrication, viii, 2, 3, 6, 21
ferrite, vii, 1, 2
ferromagnetic, 2, 19
fiber(s), 7, 8, 14, 15, 16

## G

geometry, 31, 32, 33, 34, 35, 36, 37, 39, 47
graphite, viii, 2, 3, 10, 15, 16, 21

## H

heat loss, 44
hydrogen, 30

## I

interface, ix, 13, 19, 53, 67, 68, 69, 70, 71, 72, 73, 75, 76, 79, 80, 82, 83, 85, 86
interference, 2, 31, 51

## L

layered structures, 30

# Index

leakage, 32, 45, 46
light, viii, 29, 34, 38, 47, 51, 53
light transmission, 51, 53
lithium, 18

## M

magnetic materials, vii, 2, 3, 7, 21
magnetization, 3
Maxwell equations, 68, 84, 90
microwave absorbing materials, 16
microwave absorption properties, v, vii, 1, 2, 3, 7, 8, 9, 10, 12, 13, 14, 15, 16, 18, 19, 20, 21, 22
microwave radiation, 48

## N

nanocomposites, 9, 10, 12, 13, 18, 19, 20
nanofibers, 7, 8
nanorods, 11, 12

## O

optimization, viii, 55, 61
orthogonality, 35

## P

photonic crystal resonator, 30, 32, 53
polarization, 6, 7, 8, 9, 12, 14, 15, 17, 18, 19, 20, 21, 30, 34, 37, 38, 40, 41, 43, 45, 47, 48, 51, 79, 85, 86
polypropylene, viii, 29, 32, 33, 35
polyvinylidene fluoride, 21

## Q

quantum mechanics, 30, 39

## R

radiation, vii, viii, 1, 2, 29, 32, 35, 41, 43, 44, 45, 47, 49, 52, 56, 63
reflection loss, viii, 2, 3, 8

## S

scattering, v, vii, ix, 7, 14, 18, 19, 20, 21, 44, 67, 68, 69, 70, 71, 73, 75, 77, 79, 81, 83, 85, 86, 87, 88, 89, 90, 91, 92, 93, 94, 95, 96, 97
semiconductor, 44, 52
silicon, viii, 29, 32, 33, 35, 36, 40, 44, 50, 53
$SiO_2$, 17, 18, 20
spectroscopy, vii, viii, 29, 30, 33, 52
synergistic effect, 10, 11, 13, 18, 19, 20

## T

ternary composites, 3, 14, 15, 16, 17, 18, 20, 21
THz spectroscopy, 30, 33
THz waves, 30

## U

UV radiation, 56

## V

velocity, 50

## W

wave number, 80, 85, 87
wave vector, 34
windows, 30, 31, 35, 36, 37, 38, 41, 46, 50, 51
wireless devices, vii, 1
wireless internet, 55
wireless networks, 58, 61